Heat Transfer Processes in Oscillatory Flow Conditions

Special Issue Editor
Artur J. Jaworski

MDPI • Basel • Beijing • Wuhan • Barcelona • Belgrade

Special Issue Editor
Artur J. Jaworski
University of Huddersfield
UK

Editorial Office
MDPI AG
St. Alban-Anlage 66
Basel, Switzerland

This edition is a reprint of the Special Issue published online in the open access journal *Applied Sciences* (ISSN 2076-3417) in 2017 (available at: http://www.mdpi.com/journal/applsci/special_issues/heat_transfer).

For citation purposes, cite each article independently as indicated on the article page online and as indicated below:

Lastname, F.M.; Lastname, F.M. Article title. *Journal Name* **Year**. Article number, page range.

First Edition 2018

ISBN 978-3-03842-710-0 (Pbk)
ISBN 978-3-03842-709-4 (PDF)

Cover photo courtesy of Artur J. Jaworski

Articles in this volume are Open Access and distributed under the Creative Commons Attribution license (CC BY), which allows users to download, copy and build upon published articles even for commercial purposes, as long as the author and publisher are properly credited, which ensures maximum dissemination and a wider impact of our publications. The book taken as a whole is © 2018 MDPI, Basel, Switzerland, distributed under the terms and conditions of the Creative Commons license CC BY-NC-ND (http://creativecommons.org/licenses/by-nc-nd/4.0/).

Table of Contents

About the Special Issue Editor

Professor Artur J. Jaworski, PhD; MSc(Eng); DIC; CEng; FRAeS; FHEA, is currently Chair in Mechanical Engineering at the School of Computing and Engineering, University of Huddersfield, UK. He has previously served as Chair in Energy Technology and Environment at the University of Leeds (2013–2017), Chair in Engineering and Head of Thermofluids Research Group at the University of Leicester (2011–2013) and Lecturer and Senior Lecturer at the University of Manchester (2000–2011). In 2015, he has held a Visiting Professor appointment at the Faculty of Engineering, University of Cagliari, Italy, and in 2017 a Visiting Professor appointment at the School of Computing and Engineering, University of Huddersfield, UK.

Professor Jaworski received his MSc(Eng) from the Faculty of Power and Aeronautical Engineering of the Warsaw University of Technology (1986–1991), and PhD and DIC from the Department of Aeronautics, Imperial College of Science, Technology and Medicine, London (1992–1996). He subsequently trained as postdoctoral research associate at the Department of Chemical Engineering, University of Manchester Institute of Science and Technology (1996–2000).

Professor Jaworski's research track record includes: theoretical and numerical analysis of heat and mass transfer processes in thermal-solar systems, experimental fluid dynamics and aerodynamics related to vortex dynamics and coherent structures generated by vortex breakdown flow field, aerodynamic flow control using synthetic jet actuators, sensor design, measurement and instrumentation for multiphase processes, non-invasive imaging techniques such as industrial process tomography and most recently Thermoacoustic Technologies which utilize thermal-fluid interactions between acoustic field and compressible fluid to engineer thermodynamic machines (engines and coolers) with no moving parts. He has held two prestigious fellowships in Thermoacoustic Technologies: EPSRC Advanced Research Fellowship (2004–2009) and Royal Society Industry Fellowship (2012–2015). This research gave rise to his interest in thermal—fluid processes in oscillatory flows which underpins the current Special Issue in Applied Sciences. He has co-authored in excess of 140 publications, of which around 60 are in refereed scientific journals.

Editorial

Editorial for Special Issue: "Heat Transfer Processes in Oscillatory Flow Conditions"

Artur J. Jaworski

School of Computing and Engineering, University of Huddersfield, Queensgate, Huddersfield HD1 3DH, UK; a.jaworski@hud.ac.uk

Received: 22 September 2017; Accepted: 25 September 2017; Published: 26 September 2017

Heat exchange processes in steady flows have been studied extensively over the last two hundred years, and are now part of undergraduate syllabi of most engineering courses. However, heat transfer processes in oscillatory flow conditions are still not very well understood. Their importance is well recognized in applications including Stirling machines, thermoacoustic engines, and refrigerators or pulsed-tube coolers in cryogenics. Additionally, the enhancement of heat transfer by using oscillatory, and, in some cases, pulsating flows is important in many areas of mechanical and chemical engineering for the intensification of heat transfer processes and possible miniaturization of heat exchangers of the future.

This special issue was intended as a dissemination platform for researchers working in the field to give an opportunity to consolidate the recent advances in this important research field. All types of research approaches were invited, including experimental, theoretical, computational fluid dynamics (CFD), and their mixtures, while the approaches could be both of a fundamental and applied nature. The heat transfer phenomena could be analysed from both the global (macroscopic) perspective, for instance whole heat exchanger units working in an oscillatory flow regime, and the local (microscopic) perspective, for instance the fundamentals of heat transfer in individual channels where such processes occur in an oscillatory flow regime. Invitation was also open to all disciplines, including, but not limited to, engineering, physics and chemistry, and biological and medical sciences—the underlying theme being simply heat transfer processes in oscillatory flow conditions.

The response to this invitation was very impressive—we received over 20 manuscripts. A strict refereeing process adopted by *Applied Sciences* meant that only 10 papers made it to the final special issue. Also, in the process, some papers were judged not suitable for the special issue, but generally of suitable standard for the regular issues of *Applied Sciences* and these were transferred over to alternative editors and subsequently published elsewhere. Publishing this special issue was of course a team effort, and thanks are due to all involved, in particular a group of reviewers who have to remain anonymous, my colleagues from the editorial board: Professor Yulong Ding (University of Birmingham, Birmingham, UK) and Professor Vitalyi Gusev (Université du Maine, Le Mans, France), and the tireless editorial team led by Ms. Xiaoyan Chen, without whom the special issue would have never succeeded.

Clearly, the majority of papers in this special issue (eight out of 10) deal with heat exchange processes in the context of Stirling, thermoacoustic, and pulse tube devices. These could be (somewhat arbitrarily) divided into a few themes:

The first theme is related to experimental investigations of the performance of heat exchangers in oscillatory flow conditions with the following papers:

- A Comprehensive Empirical Correlation for Finned Heat Exchangers with Parallel Plates Working in Oscillating Flow by Huang et al. [1]

- Influence of the Water-Cooled Heat Exchanger on the Performance of a Pulse Tube Refrigerator by Wang et al. [2]
- Comparative Performance of Thermoacoustic Heat Exchangers with Different Pore Geometries in Oscillatory Flow. Implementation of Experimental Techniques by Piccolo et al. [3]

The common denominator here is an attempt to quantify the heat exchanger performance by means of either a criterial equation or comparisons between various heat exchanger designs and/or geometries from the point of view of their usefulness to the overall system. A second theme represents an extension of such heat exchanger studies by the application of CFD methods, which look at the physics of the underlying processes taking place in the heat exchanger vicinity. Here, two representative papers are:

- The Effect of Temperature Field on Low Amplitude Oscillatory Flow within a Parallel-Plate Heat Exchanger in a Standing Wave Thermoacoustic System by Mohd Saat and Jaworski [4]
- Numerical Predictions of Early Stage Turbulence in Oscillatory Flow across Parallel-Plate Heat Exchangers of a Thermoacoustic System by Mohd Saat and Jaworski [5]

In addition to the above studies of heat exchangers, the third theme of papers looks at the thermal-fluid processes in the regenerator materials. This includes the experimental approaches as well as simplified one-dimensional and more complex two-dimensional CFD approaches, represented respectively by the following papers:

- Measurement of Heat Flow Transmitted through a Stacked-Screen Regenerator of Thermoacoustic Engine by Hsu and Biwa [6]
- Modeling of Heat Transfer and Oscillating Flow in the Regenerator of a Pulse Tube Cryocooler Operating at 50 Hz by Liu et al. [7]
- Friction Factor Correlation for Regenerator Working in a Travelling-Wave Thermoacoustic System by Mohd Saat and Jaworski [8]

Last, but not least, two papers were outside the area of "thermodynamic machines" described above, but rather focused on the fundamental (and mostly theoretical) studies in the general chemical engineering context where oscillatory phenomena coupled with heat transfer processes play a vital role. These two papers include:

- Excitation of Surface Waves Due to Thermocapillary Effects on a Stably Stratified Fluid Layer by Zimmerman and Rees [9]
- Heat Transfer Investigation of the Unsteady Thin Film Flow of Williamson Fluid Past an Inclined and Oscillating Moving Plate by Gul et al. [10]

The guest editor and the editorial team of *Applied Sciences* hope that the readership will find the selection of articles presented here a useful contribution to the emerging field of heat transfer processes in oscillatory flow conditions.

Conflicts of Interest: The author declares no conflicts of interest.

References

1. Huang, J.; Liu, M.; Jin, T. A Comprehensive Empirical Correlation for Finned Heat Exchangers with Parallel Plates Working in Oscillating Flow. *Appl. Sci.* **2017**, *7*, 117. [CrossRef]
2. Wang, W.; Hu, J.; Xu, J.; Zhang, L.; Luo, E. Influence of the Water-Cooled Heat Exchanger on the Performance of a Pulse Tube Refrigerator. *Appl. Sci.* **2017**, *7*, 229. [CrossRef]
3. Piccolo, A.; Siclari, R.; Rando, F.; Cannistraro, M. Comparative Performance of Thermoacoustic Heat Exchangers with Different Pore Geometries in Oscillatory Flow. Implementation of Experimental Techniques. *Appl. Sci.* **2017**, *7*, 784. [CrossRef]

4. Mohd Saat, F.A.Z.; Jaworski, A.J. The Effect of Temperature Field on Low Amplitude Oscillatory Flow within a Parallel-Plate Heat Exchanger in a Standing Wave Thermoacoustic System. *Appl. Sci.* **2017**, *7*, 417. [CrossRef]
5. Mohd Saat, F.A.Z.; Jaworski, A.J. Numerical Predictions of Early Stage Turbulence in Oscillatory Flow across Parallel-Plate Heat Exchangers of a Thermoacoustic System. *Appl. Sci.* **2017**, *7*, 673. [CrossRef]
6. Hsu, S.H.; Biwa, T. Measurement of Heat Flow Transmitted through a Stacked-Screen Regenerator of Thermoacoustic Engine. *Appl. Sci.* **2017**, *7*, 303. [CrossRef]
7. Liu, X.; Chen, C.; Huang, Q.; Wang, S.; Hou, Y.; Chen, L. Modeling of Heat Transfer and Oscillating Flow in the Regenerator of a Pulse Tube Cryocooler Operating at 50 Hz. *Appl. Sci.* **2017**, *7*, 553. [CrossRef]
8. Mohd Saat, F.A.Z.; Jaworski, A.J. Friction Factor Correlation for Regenerator Working in a Travelling-Wave Thermoacoustic System. *Appl. Sci.* **2017**, *7*, 253. [CrossRef]
9. Zimmerman, W.B.; Rees, J.M. Excitation of Surface Waves Due to Thermocapillary Effects on a Stably Stratified Fluid Layer. *Appl. Sci.* **2017**, *7*, 392. [CrossRef]
10. Gul, T.; Khan, A.S.; Islam, S.; Alqahtani, A.M.; Khan, I.; Alshomrani, A.S.; Alzahrani, A.K.; Muradullah. Heat Transfer Investigation of the Unsteady Thin Film Flow of Williamson Fluid Past an Inclined and Oscillating Moving Plate. *Appl. Sci.* **2017**, *7*, 369. [CrossRef]

© 2017 by the author. Licensee MDPI, Basel, Switzerland. This article is an open access article distributed under the terms and conditions of the Creative Commons Attribution (CC BY) license (http://creativecommons.org/licenses/by/4.0/).

Article

A Comprehensive Empirical Correlation for Finned Heat Exchangers with Parallel Plates Working in Oscillating Flow

Jiale Huang [1,2], Mianli Liu [1,2] and Tao Jin [1,2,*]

1 Institute of Refrigeration and Cryogenics, Zhejiang University, Hangzhou 310027, China; huangjiale@zju.edu.cn (J.H.); dallylau@zju.edu.cn (M.L.)
2 Key Laboratory of Refrigeration and Cryogenic Technology of Zhejiang Province, Hangzhou 310027, China
* Correspondence: jintao@zju.edu.cn; Tel.: +86-571-8795-3233

Academic Editor: Artur J. Jaworski
Received: 20 November 2016; Accepted: 13 January 2017; Published: 8 February 2017

Abstract: The oscillating-flow heat transfer performance in finned heat exchangers is one of the main factors affecting the working efficiency of regenerative heat engines and refrigerators. In addition to the working parameters, the geometrical parameters of finned heat exchangers are also major influencing factors. In the present study, the ratio of the heat exchanger length and hydraulic diameter is applied as an independent similarity criterion. An experimental study has been carried out with six different geometrical dimensions of finned heat exchangers with parallel plates, in order to analyze the impacts of fin length, plate spacing, and corresponding relative fluid displacement amplitude, under various working conditions. Based on 298 tested points, a comprehensive empirical correlation for the finned heat exchangers with parallel plates working in oscillating flow has been proposed, providing a relatively accurate prediction, with 98.6% of data in the ±20% deviation and 83.9% of data in the ±10% deviation, within the range discussed.

Keywords: oscillating flow; heat transfer; finned heat exchanger; empirical correlation

1. Introduction

Regenerative thermal engines have various advantages, such as high efficiency, low noise production, and small vibrations, which enable wide application prospects in power engineering and energy utilization. One of the crucial components of a regenerative thermal engine is the heat exchanger working in oscillating flow [1]. The heat transfer performance of the heat exchanger directly determines the efficiency of the whole system. In order to improve the thermal efficiency of the oscillating-flow heat exchanger, it is necessary to study the dynamic heat transfer characteristics and the factors influencing the gas flow in heat exchangers. The optimization of the geometrical structure and working parameters of the oscillating-flow heat exchangers, can provide theoretical guidance for the design of highly efficient regenerative thermal engines.

According to the significant differences between the oscillating and steady flows, researchers have proposed a variety of methods for characterizing the heat transfer of an oscillating-flow heat exchanger. In 1967, Richardson [2] proposed an experimental correlation of steady-state flow in a cycle, in order to approximate the characterization of oscillating flow heat transfer, which is called the time-averaged steady flow equivalent (TASFE) model. Based on thermoacoustic theory, Nika et al. [3] mathematically developed the Nusselt number characterization method in the complex field. In 1986, Gedeon [4] was the first to verify the rationality of using averaged parameters of space and cycle in the momentum and energy equations. This method doesn't require an accurate measurement of the phase difference between the heat transfer temperature gap and the heat flow. In 1996, Zhao and Cheng [5] studied the heat transfer characteristics of oscillating gas in a circular copper tube heated by constant heat flux. The average Nusselt number was correlated with the dimensionless amplitude and frequency, obtaining a preliminary empirical correlation. In 2007, Nsofor et al. [6] carried out a study on the heat transfer characteristics of the oscillating-flow heat exchanger, measuring the heat transfer in a finned-tube heat exchanger. Jaworski and Piccolo [7] obtained the temperature and velocity fields in a parallel plate channel heat exchanger, using the planar laser-induced fluorescence (PLIF) technique, combined with the particle image velocimetry (PIV). The relationship between the Nusselt number and the Reynolds number was thus obtained. Recently, Jaworski's group investigated the effect of fin length and fin spacing on the thermal performance of finned-tube heat exchangers [8]. The heat transfer rate can be predicted using the correlation between the heat transfer effectiveness, and the normalized fin spacing and fin length. In the past decade, researchers have been approaching a deeper understanding of the oscillating flow behaviors, using state-of-the-art techniques [9–14].

Tang et al. [15,16] experimentally studied the influence of working conditions, as well as the compression-expansion effect, on the Nusselt number of finned heat exchangers. An empirical correlation of the Nusselt number, with the maximum Reynolds number and Valensi number, has been summarized. Following the analysis and optimization method, this work aims to introduce the geometrical parameters as major factors of the heat transfer performance prediction. It presents an empirical correlation for the finned heat exchanger working in oscillating flow, which comprehensively considers velocity amplitude, oscillating frequency, pressure ratio, and geometrical dimension during the heat transfer process. In the following discussion, the dimensionless parameters are defined as:

$$Re_{\max} = \frac{u_A d_h}{\nu}, \tag{1}$$

$$Va = \frac{\omega\, {d_h}^2}{\nu}, \tag{2}$$

$$Nu = \frac{h d_h}{k}, \tag{3}$$

$$PR = \frac{p_{m1} + p_{A1}}{p_{m1} - p_{A1}}, \tag{4}$$

where u_A and ω are the velocity amplitude and angular frequency of the flow, respectively. h is the mean convective heat transfer coefficient. ν and k are the kinetic viscosity and thermal conductivity of the fluid, respectively. d_h is the hydraulic diameter.

For the cross-section shape of the flow channels in fin-type heat exchangers, the geometry primarily depends on the channel length l and hydraulic diameter d_h. In the design of heat exchangers working in oscillating flow, the length direction, parallel to the direction of oscillating fluid, is constrained by the gas peak-to-peak displacement. The transverse dimension, perpendicular to the direction of oscillating fluid, is limited by the thermal penetration depth δ_k, which is defined as:

$$\delta_k = \sqrt{\frac{2k}{\omega \rho_m c_p}}, \tag{5}$$

where ρ_m and c_p are the mean density and the isobaric specific heat of the fluid, respectively.

Swift [17] pointed out that the relative displacement amplitude has an important effect on the heat transfer performance. The relative displacement amplitude A_r is the ratio of the peak-to-peak displacement value and the channel length, which is defined as follows:

$$A_r = \frac{2x_A}{l} = \frac{2u_A}{\omega l}, \tag{6}$$

where x_A is the displacement amplitude of the flow. Hofler [18] and Piccolo [19] suggested that the A_r value should be close to, or larger than, one. In fact, A_r is not an independent geometrical parameter and can be practically transformed into an expression, including the dynamic parameters, as follows:

$$A_r = \left(\frac{2Re_{\max}}{Va}\right) / \left(\frac{l}{d_h}\right) \tag{7}$$

Equation (6) indicates that the influence of geometrical parameters on heat transfer performance can be discussed in relation to the controlled working conditions. As a geometrical parameter of the heat exchanger, l/d_h can be correlated as a major factor when considering the heat transfer characteristics.

Following the work presented in Reference [15], this paper presents an experimental study using six different geometrical dimensions of finned heat exchangers with parallel plates, in order to analyze the impacts of fin length, plate spacing, and corresponding relative fluid displacement amplitude, under various working conditions. This is completed in order to propose a comprehensive empirical correlation for the finned heat exchangers with parallel plates working in oscillating flow, displaying a wider range of application.

2. Experimental Methods

In order to simulate the actual working conditions of heat exchangers in regenerative thermal engines, an experimental setup is established to measure the heat transfer characteristics of the finned heat exchangers with oscillating flow conditions. The working fluid is helium. Figure 1 presents the schematic diagram of the experimental setup, and more details can be found in [15].

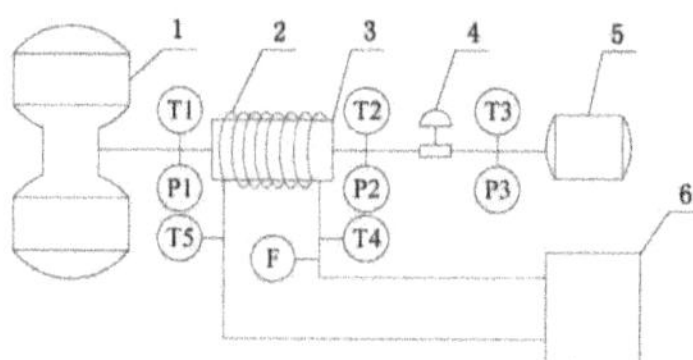

Figure 1. Schematic diagram of the experimental apparatus (1—Linear compressor; 2—Cooling water loop; 3—Testing section; 4—Adjusting valve; 5—Reservoir; 6—Thermostat).

A linear compressor (Manufactured by Lihan Thermoacoustic Technologies Co., Ltd., Shenzhen, China) is used to generate pressure wave in oscillating flow. The compressor is driven by variable-frequency power supplies so as to indirectly control the oscillating frequency of the flow by adjusting the frequency of the compressor. The input work of the compressor is controlled within 2.5 kW and the operating current is controlled below 11 A. A variable-frequency power supply allows the voltage and current limits to be set according to load characteristics required by programmable logic controllers. A frequency range of 0 Hz to 650 Hz, and a resolution of 0.01 Hz, are accessible. The combination of an adjusting valve and a gas reservoir structure is used to control and measure the velocity amplitude of the flow. More specifically, for oscillating flow, the adjusting valve and the gas reservoir can be analogous to resistance and capacity impedance, in series with the required values achieved by changing the valve opening, which controls the velocity amplitude into and out of the reservoir. By measuring the pressure wave in the gas reservoir, combined with the volume of the gas reservoir and the physical properties of the working fluid, the magnitude of the oscillating flow velocity amplitude can be determined. The heat exchanger, connected channels, and water pipes, are all thermally insulated by polystyrene foam in order to minimize the heat leakage to the surroundings.

Systematic experimental design is carried out and is based on the Box-Behnken Design (BBD), a standard method of response surface methodology (RSM) [20]. RSM is a method used to optimize the experimental conditions, suitable for solving nonlinear data processing-related problems, including techniques such as experimental design, modeling, testing the suitability of the model, searching for the best combination of conditions, and so on. Through the process of regression fitting and contour drawing, the corresponding response value of each factor's level can be easily identified. Based on the response values, the optimal response values and the corresponding experimental conditions can be obtained. At the same time, RSM can also be used to fit complex unknown function relations in a small area, with a simple first or second order polynomial model. The relatively simple calculation reduces the development cost and improves the product quality forsolving the practical problems in data processing. The software Design-Expert (Version 8.0.6, Stat-Ease Inc., Minneapolis, MN, USA) is applied in order to choose the maximum Reynolds number, the Valensi number, and l/d_h as the first-level influencing factors. After setting the corresponding names, units, and ranges, experimental operating points are suggested as a guidance for obtaining the correct experimental results, and are listed in a table with 18 typical testing points. It is proposed that these are carried out first for a quicker and better identification of each parameter's influence and trend of optimal results. In the experiments designed by RSM, the mean pressure was fixed at 3 MPa and the pressure ratio was fixed at 1.2. The maximum Reynolds number ranged from 400 to 1200 (velocity amplitude ranged from 1.0 m/s to 3.8 m/s, correspondingly), and the Valensi number ranged from 150 to 250 (working frequency ranged from 40 Hz to 90 Hz, correspondingly). With a clear understanding of each parameter's impact on the Nusselt number, indicated by the response surface, the working ranges of the system were extended to include Re_{max} values between 200 and 1200, Va values between 100 and 350, and a pressure ratio between 1.1 and 1.3.

The heat exchangers with parallel-plate channels exhibiting various l/d_h values are numbered and listed in Table 1, where Case A represents d_h = 1.5 mm and Case B represents d_h = 1.8 mm. For Case A and Case B with the same hydraulic diameter, the heat exchanger number is followed by 1, 2, or 3, to indicate a length of 15 mm, 20 mm, and 30 mm, respectively. Figure 2 shows the illustration and photo of a typical parallel-plate heat exchanger (Case A1).

Table 1. Geometrical parameters of the tested heat exchangers.

Heat Exchanger Number	l (mm)	d_h (mm)	l/d_h	Heat Exchanger Number	l (mm)	d_h (mm)	l/d_h
A1	15	1.5	10	B1	15	1.8	8.3
A2	20	1.5	13.3	B2	20	1.8	11.1
A3	30	1.5	20	B3	30	1.8	16.7

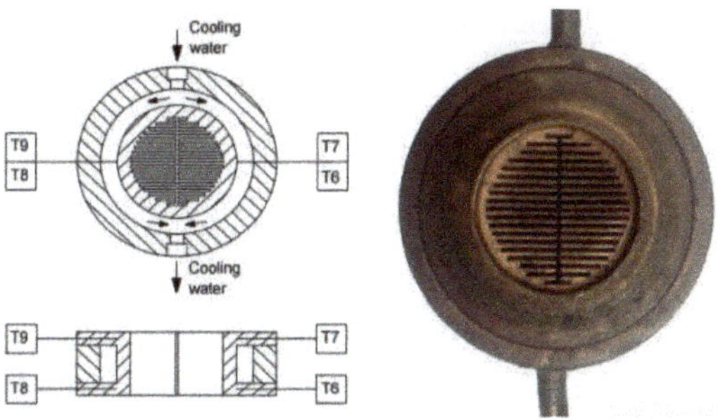

Figure 2. Illustration and photo of a typical tested heat exchanger (A1: l = 15 mm, d_h = 1.5 mm, l/d_h = 10).

Due to the heat leakage to the surroundings, the coefficient ξ is introduced into the calculation of the heat transfer rate Q_h of the gas side in the flow channel. The coefficient ξ is defined as the ratio of the actual heat transfer rate Q_h and the cooling water heat transfer rate Q_c. Q_c can be calculated as:

$$Q_c = q_v \rho_{\text{water}} c_{\text{water}} (t_2 - t_1) \tag{8}$$

where q_v is the cooling water flow rate, and ρ_{water} and c_{water} are the density and heat capacity of water, respectively. t_1 and t_2 are the inlet and outlet temperatures of the cooling water, respectively. Because of the use of the same thermal insulation method as in Ref. [15], the expression of coefficient ξ is taken as:

$$\xi = \frac{Q_h}{Q_c} = \frac{1}{1 - (Q_h - Q_c)/Q_h} = \frac{1}{1 + 5.43 \times 10^{-5} t_h^2 - 3.556 \times 10^{-2} t_h + 5.77} \tag{9}$$

where t_h is the average temperature of the cooling water.

3. Measurement Uncertainty

The experimental results are finally characterized as the average Nusselt number, which is actually a function of nine measured parameters. The Nusselt number is calculated using the following expression:

$$Nu = \frac{hd_h}{k} = \frac{\xi Q_c d_h}{kA_s\left(t_f - t_{wi}\right)} = \frac{\xi d_h c_{\text{water}} \rho_{\text{water}} (t_2 - t_1) q_v}{kA_s\left(t_f - t_{wi}\right)} \tag{10}$$

where A_s is the calculated effective heat transfer area for each finned heat exchanger; t_f and t_{wi} are the mean temperatures of the fluid and wall on the internal side, respectively.

Since the measurement errors of different variables are independent of each other, according to the principle of error propagation, the uncertainty of the Nusselt number is:

$$\sigma_{Nu} = \sqrt{\sum_{i=1}^{9} \left(\frac{\partial f}{\partial x_i}\right)^2 \sigma_{x_i}{}^2} \tag{11}$$

where $\partial f / \partial x_i$ is the partial derivative of the function on one of the measured parameters. σ_{xi} is the uncertainty of each measured parameter. For simplicity, the influence of five measurement errors on the Nusselt number were considered, including the inlet and outlet cooling water temperatures t_1 and t_2, the fluid mean temperature t_f, the mean wall temperature t_{wo} on the external side, and the cooling water flow rate q_v. According to the principle of error propagation:

$$\begin{aligned}
\left(\tfrac{\partial f}{\partial x_i}\right)^2 &= \left(\tfrac{\partial Nu}{\partial q_v}\right)^2 = \left(\tfrac{\xi d_h c_{\text{water}} (t_2 - t_1)}{kA_s (t_f - t_{wo})}\right)^2 \\
\left(\tfrac{\partial f}{\partial x_i}\right)^2 &= \left(\tfrac{\partial Nu}{\partial q_v}\right)^2 = \left(\tfrac{\xi d_h c_{\text{water}} (t_2 - t_1)}{kA_s (t_f - t_{wo})}\right)^2 \\
\left(\tfrac{\partial f}{\partial x_i}\right)^2 &= \left(\tfrac{\partial Nu}{\partial t_f}\right)^2 = \left(\tfrac{\partial Nu}{\partial t_{wo}}\right)^2 = \left(\tfrac{\xi d_h c_{\text{water}} q_v (t_2 - t_1)}{kA_s (t_f - t_{wo})^2}\right)^2
\end{aligned} \tag{12}$$

After calculating the error propagation, the total uncertainty of the Nusselt number can be obtained, according to Equation (10).

4. Results and Discussion

Firstly, in order to producea quick sketch of self-consistency and the impact trends of parameters, the analysis of variance is carried out to show the normal distribution plot of the residuals of 18 typical tested points (shown in Figure 3), which indicates self-consistency of the results. In Figure 3, the data show the normal probability distribution of residuals after an analysis of variance (ANOVA) of the tested results. The horizontal axis stands for the internally studentized residual, which is the quotient resulting from the division of a residual by an estimate of its standard deviation. The vertical axis stands for its corresponding normal possibility, which has exponential coordinates, hence, the normal distribution plot of residuals can show its self-consistency if the data lie in a neat straight line. The maximum deviation is below 10% with the correlation fitting based on RSM. The three-dimensional response surface is plotted into a polynomial correlation, with the highest order of two. Figure 4 shows the response surface plot with a fixed maximum Reynolds number of 800 for demonstration.

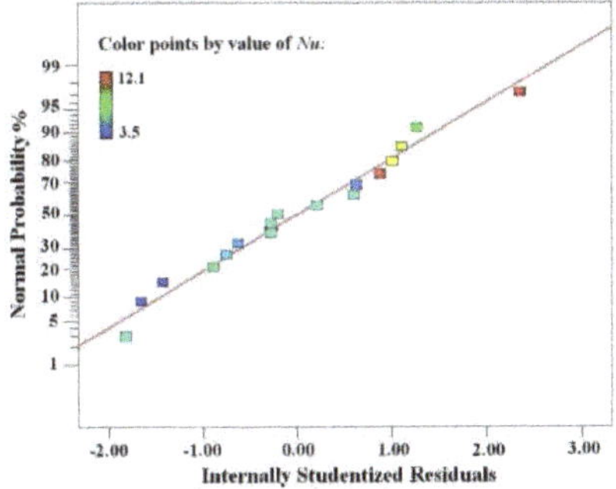

Figure 3. Normal plot of residuals for 18 typical tested points.

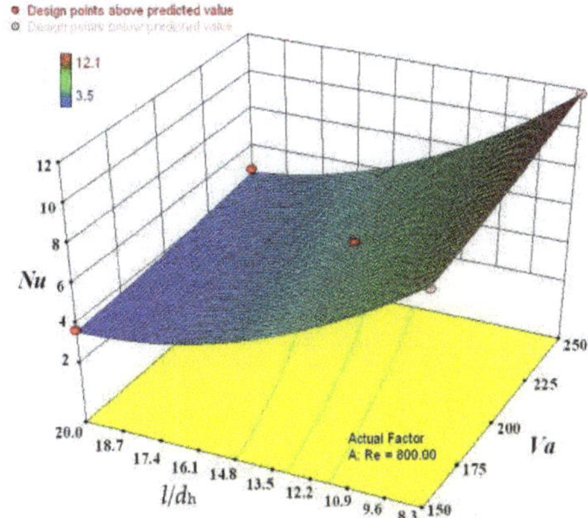

Figure 4. Response surface with a fixed maximum Reynolds number of 800.

From the response surfaces with fixed various parameters, it is noted that l/d_h, as a geometrical parameter, has a negative influence on the Nusselt number. The rise of l/d_h leads to a dramatic weakening in the heat transfer performance. Along with the fact that the maximum Reynolds number and Valensi number both have a positive impact on the Nusselt number, more detailed experimental results are obtained with error bars indicating the uncertainty of acquired data, which is not higher than 5.8%.

Figures 5–7 present the Nusselt number variations with six different l/d_h values when the Valensi numbers are 150, 200, and 250, respectively. These values are realized by the combination of three different lengths and two different hydraulic diameters, as listed in Table 1. It can be seen from the figures that, when the Reynolds number and Valensi number are both a controlled constant, the Nusselt number decreases as l/d_h is increased, reflected in the response surface illustrated in Figure 4. The oscillating-flow heat transfer performance will be significantly weakened when l/d_h is increased within the range discussed.

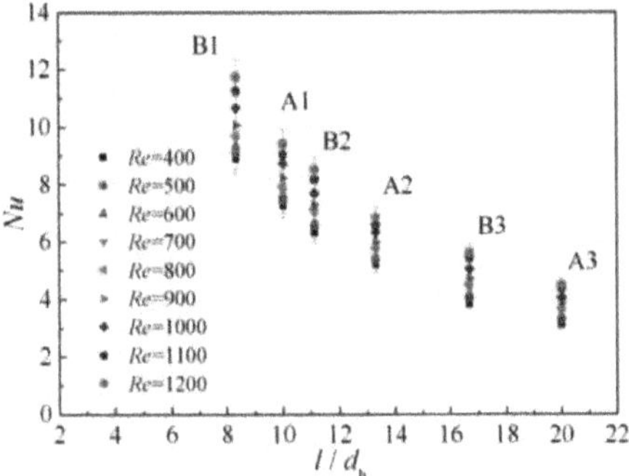

Figure 5. Nusselt number variation with l/d_h, Va = 150.

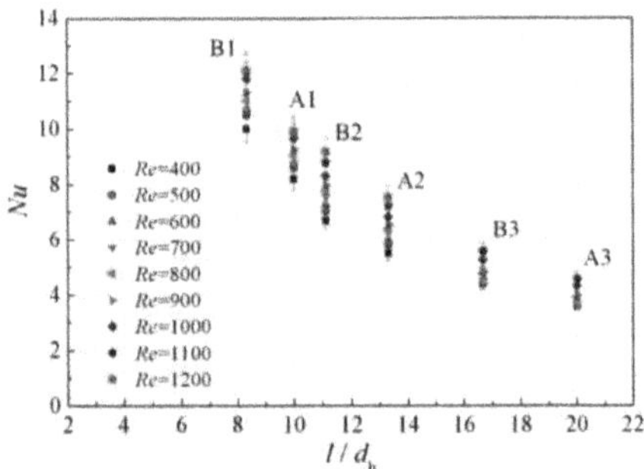

Figure 6. Nusselt number variation with l/d_h, Va = 200.

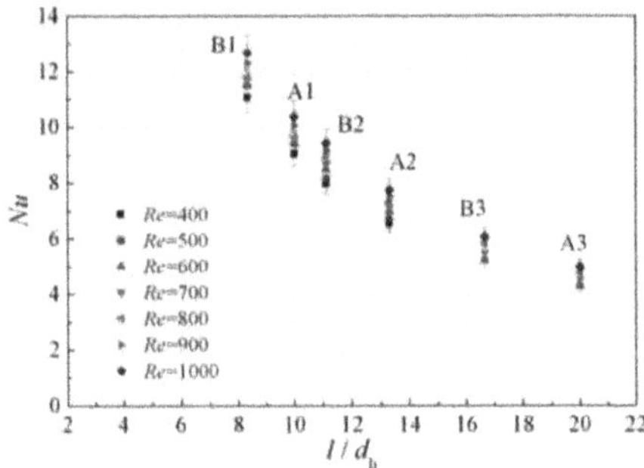

Figure 7. Nusselt number variation with l/d_h, Va = 250.

The results indicate that when the hydraulic diameter of the channel (roughly twice of the plate spacing) is constant, for the heat transfer with a smaller length (l/d_h is relatively small), the heat transfer performance of the parallel plates in oscillating helium flow is significantly better than that of heat exchangers with longer fins. It is assumed that the heat transfer process is enhanced by the entrance effect of oscillating flow. In addition, if the heat exchanger length is too large, being even longer than the peak-to-peak displacement of working fluid, part of the fluid particles will be trapped in the flow channel experiencing the reciprocating motion. As the "trapped" fluid has a temperature similar to that of the wall, the oscillating heat transfer will be relatively weak. From the view of the whole heat exchanger, it can be regarded that the central part of the heat transfer area doesn't play a good role in the heat transfer process, which will lead to a lower Nusselt number of the oscillating flow inside the entire channel. With longer channels, there will be a larger area that has no effective heat transfer. Hence, for the heat exchanger types A and B, the obtained magnitudes have orders of B1 > B2 > B3 and A1 > A2 > A3. Numerical simulation by Ishikawa et al. [21] showed that the net heat transfer rate between solid and fluid states mainly occurred on the inlet and outlet faces of the heat exchanger with parallel plates. When the fin length is increased, the heat flux on inlet and outlet faces can be positive or negative, leading to a decrease in the net heat flux. As a result, it is also possible that the heat transfer performance of the oscillating flow in the heat exchanger with a length of 15 mm, is much better than the others due to the increased heat transfer rate per unit area.

Figures 8–10 present the Nusselt number variations of A_r when Valensi numbers are 150, 200, and 250, respectively. These figures also show that for heat exchangers with the same hydraulic diameter, but different length, the Nusselt number decreases as the heat exchanger length is increased. It can be seen that when *Re* and *Va* are controlled, heat exchangers with a length of 20 mm (A2 and B2) have about a 30% lower *Nu* number than those with a length of 15 mm (A1 and B1), and heat exchangers with a length of 30 mm (A3 and B3) have about a 50% lower *Nu* number than those with a length of 15 mm. When considering those with a length of 15 mm (A1 and B1), it is evident that better heat transfer performance mainly occurs when A_r is larger than 1, supporting the conclusion offered by Hofler [18] and Piccolo [19]. In fact, when *Re* and *Va* are controlled constants, the relative fluid displacement amplitude A_r is equivalent to the reciprocal of l/d_h.

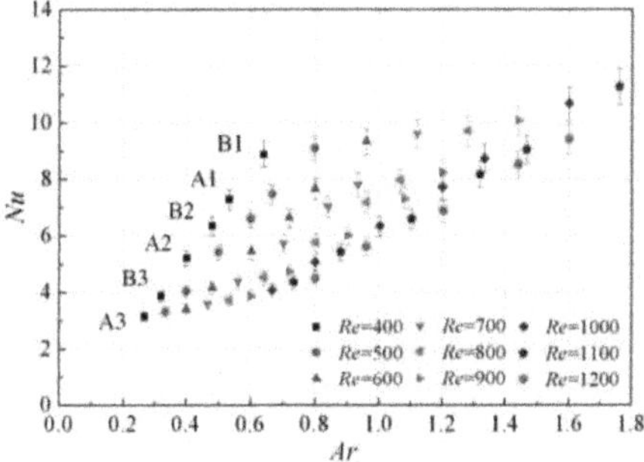

Figure 8. Nusselt number variation with A_r, *Va* = 150.

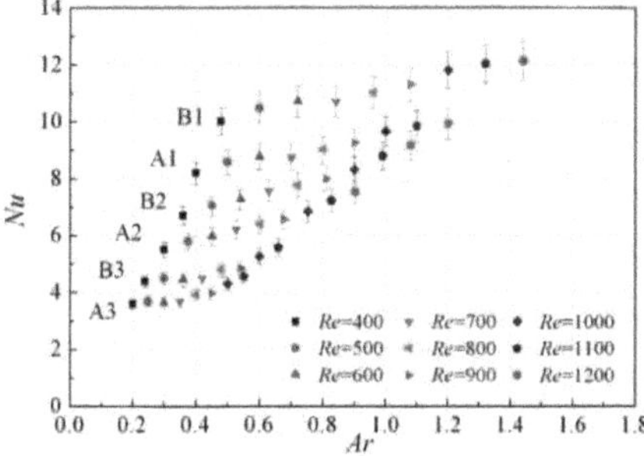

Figure 9. Nusselt number variation with A_r, *Va* = 200.

Compared to Figures 5–7, the *Nu* variations of A_r reflect the same physical pattern. However, the impact of relative fluid displacement amplitude is actually a result of comprehensive effects; an aspect which has been neglected in the previous study. Piccolo [22] numerically studied the relationship between the mean convective heat transfer coefficient *h*, and the relative displacement amplitude

A_r, under the same pressure ratio based on energy conservation. Results showed that remarkably, h increased with increasing A_r when $A_r > 1$, while h remained constant when $A_r \leq 1$, which does not reflect the results concluded in this paper. As A_r includes the geometrical parameters, as well as the impacts of *Re* and *Va*, it is difficult to explain the direct relationship between the oscillating-flow heat transfer and each parameter. In comparison, l/d_h, instead of A_r, can be chosen as an important geometrical parameter, which should be taken into account when considering the empirical correlation for the finned heat exchangers with parallel plates, working under oscillating conditions.

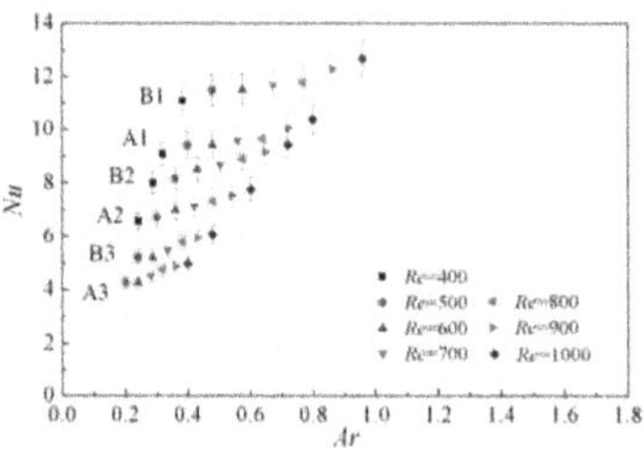

Figure 10. Nusselt number variation with A_r, *Va* = 250.

Figures 11–13 present the actual heat transfer rate variations of A_r when the Valensi numbers are 150, 200, and 250, respectively. The results show that, for the same type of heat exchanger with the same hydraulic diameter, there exists an optimal fin length for achieving the required heat load. Out of the three lengths tested, the length of 20 mm achieves the largest heat transfer rate for each working condition. In the practical design of regenerative thermal systems, the fin efficiency and the effective heat transfer area should be considered, as well as the space-cycle average Nusselt number.

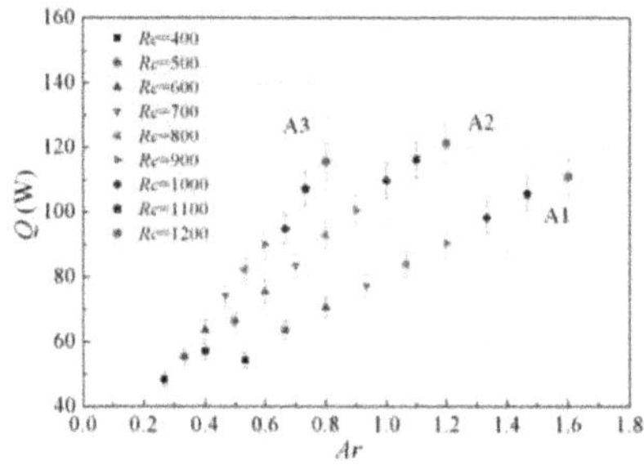

Figure 11. Heat transfer rate variation with A_r, *Va* = 150.

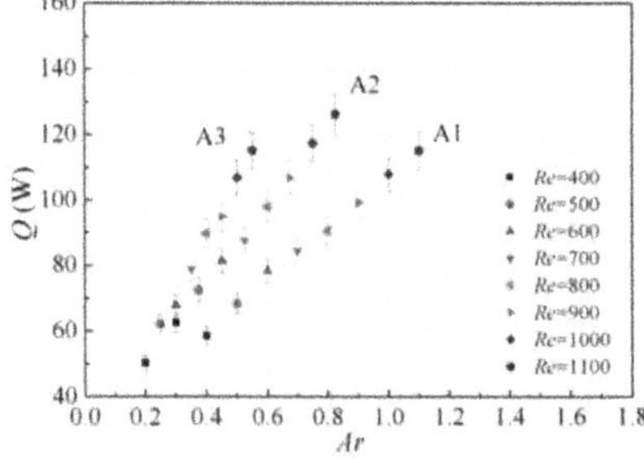

Figure 12. Heat transfer rate variation with A_r, *Va* = 200.

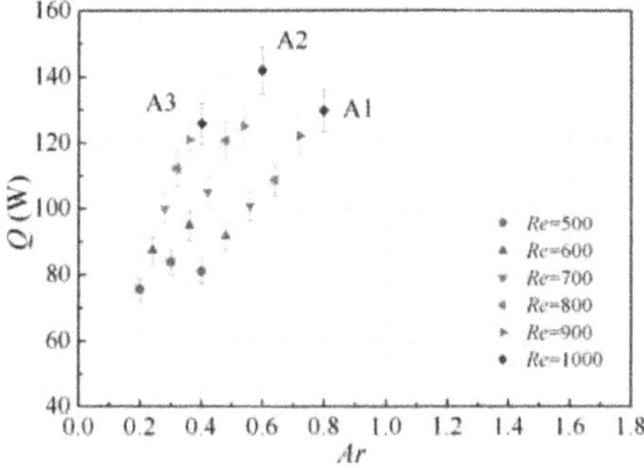

Figure 13. Heat transfer rate variation with A_r, $Va = 250$.

In order that the experimental data can be conveniently applied to the design of finned heat exchangers working under oscillating conditions in regenerative thermal engines, further data fitting was implemented to obtain a relatively comprehensive empirical correlation. The Reynolds number, Valensi number, Mach number, Prandtl number, specific heat ratio, ratio of length and hydraulic diameter, and pressure ratio *PR*, are considered as the main similarity criterions relating to the oscillating-flow heat transfer in heat exchanging channels [5]. This experiment adopted helium gas as the working fluid, so the specific heat ratio and Prandtl number can be considered as nearly constant. In addition, the Mach number implies the compressibility of fluid, which was kept at a value much lower than one in the experiment. Finally, based on the discussion above, the expression of the empirical correlation is defined as:

$$Nu = aPR^b\left(\frac{l}{d_h}\right)^c Re^m Vd^n \tag{13}$$

It is ascertained that $b = 6.138$ from the linear fitting of log(Nu) and log(PR) with a coefficient determination of $R^2 = 0.9798$, as shown in Figure 14. Following this, log($Nu/PR^{6.138}$) and log($Re_{\max}$) are linearly fitted with a slope of $m = 0.153$ and a coefficient determination of $R^2 = 0.8158$. Next, by completing log($Nu/PR^{6.138}/Re_{\max}{}^{0.153}$) and log($Va$) linear fitting with $n = 0.504$ and $R^2 = 0.8296$, the coefficient c is determined as -1.137 and a is calculated from the intercept, producing a value of 1.021. Figures 14–17 show the determination of all the coefficients.

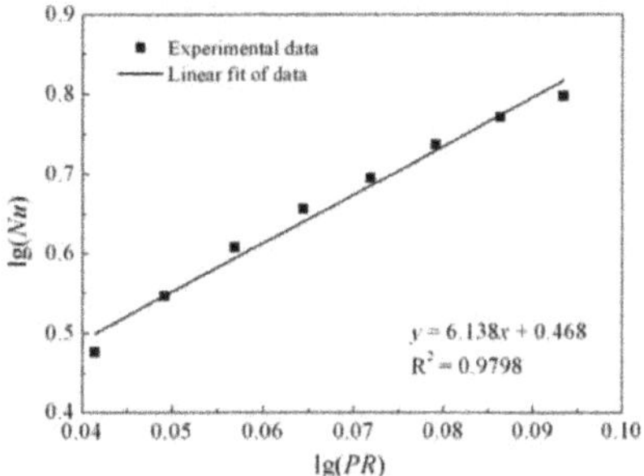

Figure 14. Determination of coefficient b.

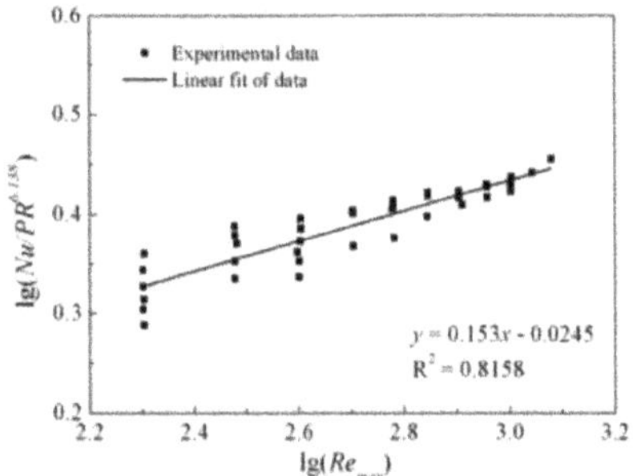

Figure 15. Determination of coefficient *m*.

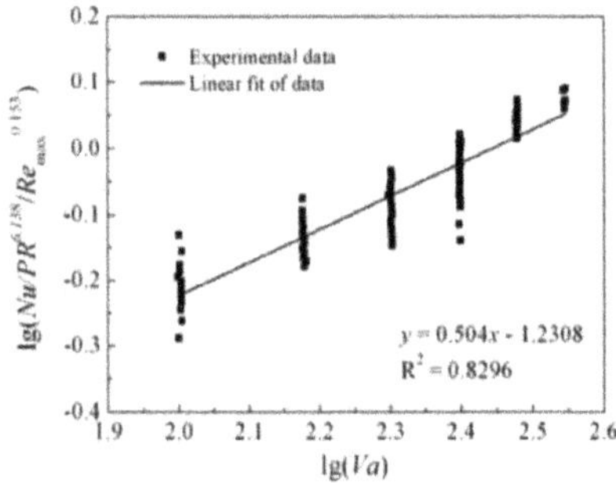

Figure 16. Determination of coefficient *n*.

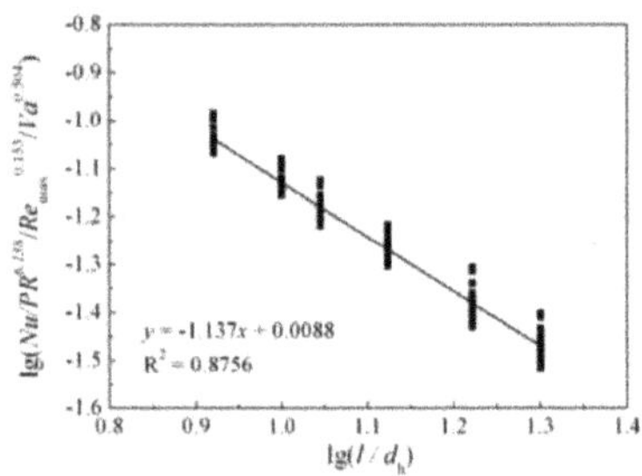

Figure 17. Determination of coefficients *c* and *a*.

As a result, within the ranges used in the experiments, i.e., $200 < Re_{\max} < 1200$, $100 < Va < 350$, $1.1 < PR < 1.3$, and $8.3 < l/d_h < 20$, the space-cycle average Nusselt number of oscillating-flow heat transfer in the finned heat exchanger with parallel plates, can be predicted using the following correlation:

$$Nu = 1.021\frac{PR^{6.138}Re^{0.153}Va^{0.504}}{(l/d_h)^{1.137}} \tag{14}$$

To prove the self-consistency of the empirical correlation, the predicted results are compared to the corresponding experimental results, as shown in Figure 18. Out of the total 298 tested data points, 98.6% of the predicted points lie within adeviation of ±20%, and 83.9% of the predicted points lie within adeviation of ±10%, indicating that the comprehensive empirical correlation is able to provide acritical guidance for the design of finned heat exchangers in regenerative thermal engines.

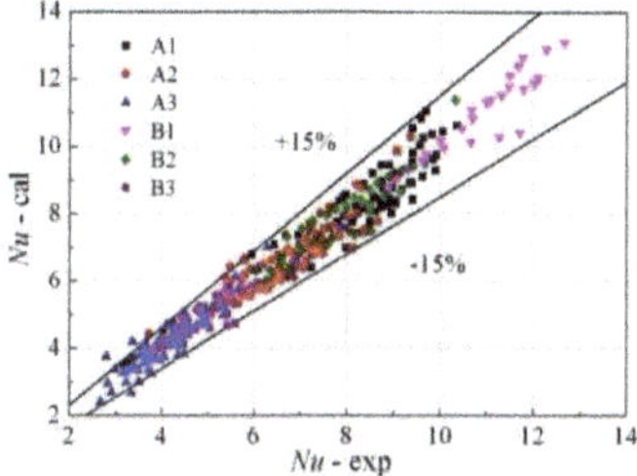

Figure 18. Comparison of calculated data with experimental data.

5. Conclusions

The space-cycle average Nusselt number is introduced to characterize the heat transfer performance of oscillating helium flow in the finned heat exchangers with parallel plates. The relationship between the Nusselt number and l/d_h, and the relative fluid displacement amplitude A_r, are analyzed based on experimental results, respectively. A decrease in l/d_h leads to a remarkable increase in the Nusselt number. Therefore, l/d_h appears to be the most appropriate choice in terms of the actual value of the heat transfer rate. An A_r value which is larger than one improves the oscillating-flow heat transfer inside the parallel plates. It is suggested that l/d_h would be a more reasonable parameter for characterizing the geometrical similarity. Based on a total number of 298 tested points, a comprehensive empirical correlation for finned heat exchangers with parallel plates working in oscillating flow, has been proposed. A total of 98.6% of the predicted points lie within adeviation of ±20%, and 83.9% of the predicted points lie within adeviation of ±10%, which indicates that the comprehensive empirical correlation is able to provide critical guidance for the design of finned heat exchangers in regenerative thermal engines.

Acknowledgments: The project is financially supported by the National Key Basic Research Program of China (No. 613322) and the National Natural Science Foundation of China (No. 51576170).

Author Contributions: Jiale Huang and Tao Jin conceived and designed the experiments; Mianli Liu performed the experiments; Jiale Huang and Mianli Liu analyzed the data; Tao Jin contributed analysis tools; Jiale Huang and Tao Jin wrote the paper.

Conflicts of Interest: The authors declare no conflict of interest.

References

1. Jin, T.; Huang, J.L.; Feng, Y.; Yang, R.; Tang, K.; Radebaugh, R. Thermoacoustic prime movers and refrigerators: Thermally powered engines without moving components. *Energy* **2015**, *93*, 828–853. [CrossRef]
2. Richardson, P.D. Effects of sound and vibration on heat transfer. *Appl. Mech. Rev.* **1967**, *20*, 201–217.
3. Nika, P.; Bailly, Y.; Guermeur, F. Thermoacoustics and related oscillatory heat and fluid flows in micro heat exchangers. *Int. J. Heat Mass Transf.* **2005**, *18*, 3773–3792. [CrossRef]
4. Gedeon, D. Mean-parameter modeling of oscillating flow. *J. Heat Transf.* **1986**, *108*, 513–518. [CrossRef]
5. Zhao, T.S.; Cheng, P. Oscillatory heat transfer in pipe subjected to a laminar reciprocating flow. *J. Heat Transf.* **1996**, *3*, 92–597. [CrossRef]
6. Nsofor, E.C.; Celik, S.; Wang, X.D. Experimental study on the heat transfer at the heat exchanger of the thermoacoustic refrigerating system. *Appl. Therm. Eng.* **2007**, *27*, 2435–2442. [CrossRef]
7. Jaworski, A.J.; Piccolo, A. Heat transfer processes in parallel-plate heat exchangers of thermoacoustic devices—Numerical and experimental approaches. *Appl. Therm. Eng.* **2012**, *42*, 145–153. [CrossRef]
8. Kamsanam, W.; Mao, X.N.; Jaworski, A.J. Thermal performance of finned-tube thermoacoustic heat exchangers in oscillatory flow conditions. *Int. J. Therm. Sci.* **2016**, *101*, 169–180. [CrossRef]
9. Jaworski, A.J.; Mao, X.N.; Mao, X.R.; Yu, Z.B. Entrance effects in the channels of the parallel plate stack in oscillatory flow conditions. *Exp. Therm. Fluid Sci.* **2009**, *33*, 495–502. [CrossRef]

10. Yu, Z.B.; Mao, X.N.; Jaworski, A.J. Experimental study of heat transfer in oscillatory gas flow inside a parallel-plate channel with imposed axial temperature gradient. *Int. J. Heat Mass Transf.* **2014**, *77*, 1023–1032. [CrossRef]
11. Kamsanam, W.; Mao, X.N.; Jaworski, A.J. Development of experimental techniques for measurement of heat transfer rates in heat exchangers in oscillatory flows. *Exp. Therm. Fluid Sci.* **2015**, *62*, 202–215. [CrossRef]
12. Mathie, R.; Markides, C.N. Heat transfer augmentation in unsteady conjugate thermal system—Part I: Semi-analytical 1-D framework. *Int. J. Heat Mass Transf.* **2013**, *1*, 802–818. [CrossRef]
13. Mathie, R.; Nakamura, H.; Markides, C.N. Heat transfer augmentation in unsteady conjugate thermal systems-Part II: Applications. *Int. J. Heat Mass Transf.* **2013**, *1*, 819–833. [CrossRef]
14. Mathie, R.; Markides, C.N.; White, A.J. A framework for the analysis of thermal losses in reciprocating compressors and expanders. *Heat Trans. Eng.* **2014**, *16–17*, 1435–1449. [CrossRef]
15. Tang, K.; Yu, J.; Jin, T.; Wang, Y.P.; Tang, W.T.; Gan, Z.H. Heat transfer of laminar oscillating flow in finned heat exchanger of pulse tube refrigerator. *Int. J. Heat Mass Transf.* **2014**, *70*, 811–818. [CrossRef]
16. Tang, K.; Yu, J.; Jin, T.; Gan, Z.H. Influence of compression-expansion effect on oscillating-flow heat transfer in a finned heat exchanger. *J. Zhejiang Univ. Sci. A* **2013**, *14*, 427–434. [CrossRef]
17. Swift, G.W. Thermoacoustic engines. *J. Acoust. Soc. Am.* **1988**, *84*, 1145–1180. [CrossRef]
18. Hofler, T.J. Effective heat transfer between a thermoacoustic heat exchanger and stack. *J. Acoust. Soc. Am.* **1993**, *94*, 1772. [CrossRef]
19. Piccolo, A. Numerical computation for parallel plate thermoacoustic heat exchangers in standing wave oscillatory flow. *Int. J. Heat Mass Transf.* **2011**, *54*, 4518–4530. [CrossRef]
20. Chiang, K.T. Modeling and optimization of designing parameters for a parallel-plain fin heat sink with confined impinging jet using the response surface methodology. *Appl. Therm. Eng.* **2007**, *27*, 2473–2482. [CrossRef]
21. Ishikawa, H.; Mee, D.J. Numerical investigation of flow and energy fields near a thermoacoustic couple. *J. Acoust. Soc. Am.* **2002**, *111*, 831–839. [CrossRef] [PubMed]
22. Piccolo, A.; Pistone, G. Estimation of heat transfer coefficients in oscillating flow: The thermoacoustic case. *Int. J. Heat Mass Transf.* **2006**, *49*, 1631–1642. [CrossRef]

© 2017 by the authors. Licensee MDPI, Basel, Switzerland. This article is an open access article distributed under the terms and conditions of the Creative Commons Attribution (CC BY) license (http://creativecommons.org/licenses/by/4.0/).

Article

Influence of the Water-Cooled Heat Exchanger on the Performance of a Pulse Tube Refrigerator

Wei Wang [1,2], Jianying Hu [1,*], Jingyuan Xu [1,2], Limin Zhang [1] and Ercang Luo [1]

1 TKey Laboratory of Cryogenics, Technical Institute of Physics and Chemistry, Chinese Academy of Sciences, Beijing 100190, China; wangwei13@mails.ucas.ac.cn (W.W.); xujingyuan@mail.ipc.ac.cn (J.X.); liminzhang@mail.ipc.ac.cn (L.Z.); ecluo@mail.ipc.ac.cn (E.L.)

2 The College of Materials Science and Opto-electronic Engineering, University of Chinese Academy of Sciences, Beijing 100049, China

* Corresponding: jyhu@mail.ipc.ac.cn; Tel.: +86-10-8254-3733

Academic Editor: Artur Jaworski
Received: 4 January 2017; Accepted: 23 February 2017; Published: 28 February 2017

Abstract: The water-cooled heat exchanger is one of the key components in a pulse tube refrigerator. Its heat exchange effectiveness directly influences the cooling performance of the refrigerator. However, effective heat exchange does not always result in a good performance, because excessively reinforced heat exchange can lead to additional flow loss. In this paper, seven different water-cooled heat exchangers were designed to explore the best configuration for a large-capacity pulse tube refrigerator. Results indicated that the heat exchanger invented by Hu always offered a better performance than that of finned and traditional shell-tube types. For a refrigerator with a working frequency of 50 Hz, the best hydraulic diameter is less than 1 mm.

Keywords: heat exchanger; pulse tube refrigerator; shell-tube types

1. Introduction

Owing to their high reliability, high efficiency, and compact size, pulse tube refrigerators have attracted much attention in the past decades [1]. Many of them have been developed to cool devices such as infrared detectors, HTS filters, and germanium detectors. Their cooling power is often less than ten watts. Recent advances in high-temperature superconductivity (HTS) devices, small gas liquefaction, and cryogenic storage tanks, have spurred the demand for high-capacity refrigerators. The cooling power required by such applications varies from hundreds to thousands of watts, and the corresponding cooling temperature varies from 80 to 120 K [2–4]. The pulse tube refrigerator was considered to be one of the best candidates, so the development and use of high-capacity pulse tube refrigerators boomed. At present, the cooling power of these refrigerators can exceed 1 kW at a temperature of 80 K, and a relative Carnot efficiency of 22% has been achieved [5–9]. Although this efficiency is quite high, some studies have indicated that the room temperature heat exchanger could be further improved [10–12].

The water-cooled heat exchanger (WCHX) is one of the core components in a pulse-tube cooler [13,14]. The function of the heat exchanger is to transfer time-averaged heat from the gas to the water, which acts as an external heat sink. It must provide good thermal contact between the two flowing streams, while causing a minimal pressure drop in either stream. There are three typical configurations for water-cooled heat exchangers employed in pulse tube refrigerators. The first is the finned-tube type introduced in References [15,16]. In this type, fins of different lengths are arranged in a round case at regular intervals. Thermal contact between the fins and tubes acting as channels for the water is ensured by soldering. Although this is a very appropriate high-capacity design in theory, it is technically difficult. The water-cooled heat exchanger presented in References [17–19], shown in

Figure 1b, can also be considered a fin-tube type. The fins are machined by wire electrical discharge machining (WEDM), and the water only flows around the circumference of the case. Meanwhile, the temperature difference on the fins caused by heat conductance will increase with the size of the exchanger [8,20]. It is much easier to machine, but more expensive. The third configuration is the shell-tube type, shown in Figure 1c. As is known, the heat transfer coefficient at the water side is much higher than that at the gas side [21], so the heat exchange at the gas side should be strengthened. With the present machining process capability, it is impossible to produce such fins in a pipe with a small hydraulic diameter. If the hydraulic diameter is less than 1 mm, the flow area of the gas is often less than 15% of the cross-sectional area of the heat exchanger. To overcome this drawback of the shell-tube heat exchanger, Hu introduced another configuration, as shown in Figure 1d [22]. Small tubes acting as fins are welded inside the larger tubes, to strengthen the heat transfer at the gas side. Because the fins are not long, a uniform temperature distribution on the heat exchanger can be ensured. The hydraulic diameter and flow area can be easily adjusted by changing the parameters of the small tubes. The effectiveness of this configuration was confirmed in Reference [19].

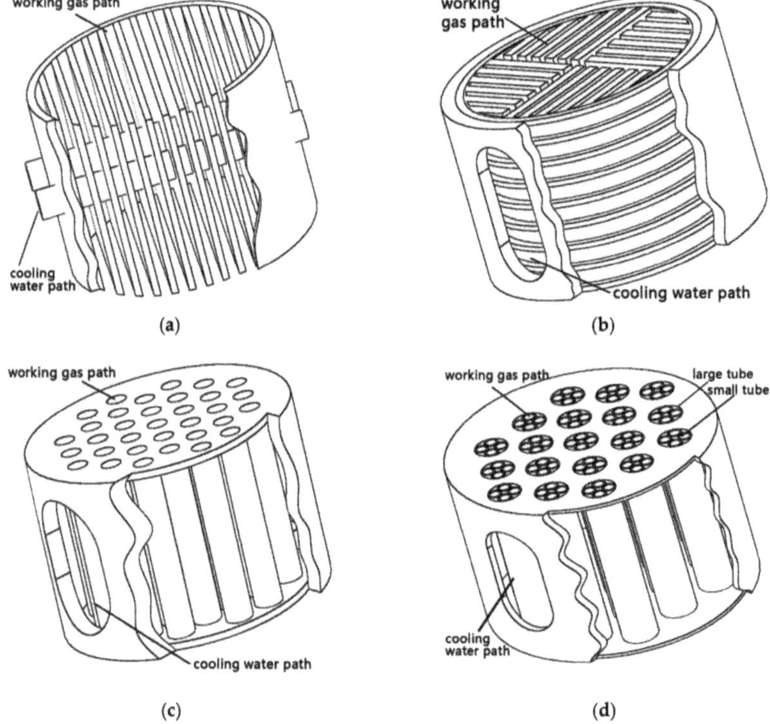

Figure 1. Four configurations of WHCXs. (**a**) Finned-tube type; (**b**) WEDM machined type which can be considered as a finned-tube type; (**c**) shell-tube type; (**d**) new type introduced by Hu.

Generally, a smaller hydraulic diameter and high flow velocity can improve heat transfer, but also result in greater flow loss. A good design achieves a reasonable compromise between these variables, in order to obtain the best performance of the refrigerator. The pressure drop can be calculated based on thermoacoustic theory. Meanwhile, the heat transfer is hard to evaluate because the empirical correlation for steady flow cannot be directly applied. A well accepted heat transfer correlation for oscillating flow is still absent, although much investigation has been carried out in this area [13,18,23,24]. Thus, in this work, we used experiments to determine the influence of factors such as the hydraulic diameter, length, contact thermal resistance, and porosity, on the cooling performance of a pulse tube refrigerator.

2. Experimental Apparatus and WCHXs

2.1. Experimental Apparatus

The experimental apparatus is schematically shown in Figure 2. It mainly consisted of a linear compressor and an in-line pulse-tube cooler. The main geometric parameters of the cooler and its operating conditions are listed in Table 1. The water-cooled heat exchanger was replaceable, so that different heat exchangers could be tested. The heat exchangers were cooled by circulating water with a temperature of 293 K. Because the hydraulic diameter of the heat exchanger was quite small, it was difficult to place thermometers inside it, in order to measure the gas or the wall temperatures. Subsequently, only a calibrated K-type sheathed thermocouple (T_0) was installed close to the inlet of the heat exchanger, to measure the gas temperature.

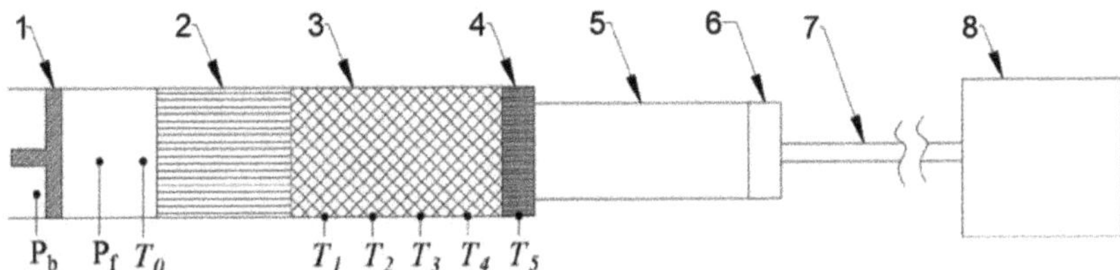

Figure 2. Schematic of the experimental apparatus. 1 Liner compressor; 2 WCHX; 3 Regenerator; 4 Clod head; 5 Pulse-tube; 6 Flow straightener; 7 Inertance tube; 8 Gas reservoir.

The regenerator housing was filled with a 300-mesh stainless-steel screen with a porosity of 0.7. Four calibrated PT-100 resistance thermometers (T_1–T_4) with an accuracy of ±0.1 K were placed on the outer wall of the regenerator, distributed with equal spacing. If the wall temperature equals the screen temperature in the same cross-section, then the temperature in the regenerator can be obtained from the four thermometers. The temperature of the cold head was measured with another PT-100 (T_5). Three constantan wires heated by a direct voltage source were mounted on the cold head, to simulate cooled loads. The voltage was adjusted to keep the temperature of the cold head at 80 K. The heat power was read from the direct voltage source, which had an accuracy of ±0.3%. The regenerator, cold head, and pulse tube, were enclosed in a vacuum chamber. The flow straightener was made of 5-mm long 40-mesh copper screens. The inertance tube was cooled with a water jacket. The linear compressor was used to provide acoustic power for the pulse-tube cooler. The efficiency of the compressor was not important in this study, so the match between the compressor and the refrigerator was not optimized. The output acoustic power of the compressor was calculated from the pressures (P_f and P_b, respectively) in its front and back spaces. P_f and P_b were measured using high-precision dynamic pressure sensors, supplied by PCB Piezotronics shown in Figure 2. The input acoustic power, W_a, was calculated as:

$$W_a = \frac{1}{2}\left|P_f\right|\left|\frac{i\omega V_b}{\gamma P_0}P_b\right|\cos\theta_{PU} \tag{1}$$

where θ_{PU} is the phase difference between the pressure wave and the volume flow rate, ω is the angular frequency, V_b is the back space volume of the compressor, P_0 is the mean pressure, and γ is the specific heat ratio [25]. The accuracy of the pressure sensor measuring P_0 was 0.2%. The non-linearity of the pressure sensors measuring P_f and P_b was 1%.

Table 1. Operating conditions and component size.

Item	Item	Value
Operating condition	Working gas	helium
	Operating pressure	3 MPa
	Frequency	55 Hz
	Cooling temperature	80 K
Component size	Regenerator	75 × 70 (300 mesh)
	Pulse tube	37 × 150
	Cold head	75 × 30
	Inertance tubes	10 × 2300

All listed dimensions are inner diameter × length (all in mm).

2.2. Water-Cooled Heat Exchangers

Figure 3 shows cross-sectional photographs of the WHCXs used in the experiment. The details of each WHCX are presented in Table 2. WCHX 1 was machined by WEDM. It can be considered a finned-tube type. WCHX 2 had a traditional shell-tube structure, while WCHXs 3–6 were the new shell-tube types introduced by Hu. In WCHX 3, the small tubes acting as fins in the large tubes were larger than those in WCHX 4, so were deformed during processing. They were more triangular than circular. WCHX 5 and 6 had the same gas channels as WHCX 4, so their photographs are not shown in Figure 3. WCHX 4 and WCHX 5 had the same parameters. The only difference was that in WCHX 4, the small tubes were welded to the large tubes (before processing, the welder was electroplated to the outer surface of the small tubes). In WCHX, the small and large tubes were just touching, without welding (Please refer to Reference [14] for the processing details). To remove the influence of different gas velocities, the porosity (area of gas channel to cross-sectional area) was maintained at about 25%. With this porosity, the gas velocity varied from 10 to 29 m/s, when the input acoustic power was changed from 1280 to 5000 W. To investigate the influence of porosity, some channels were blocked with thin round plates in WCHX 7.

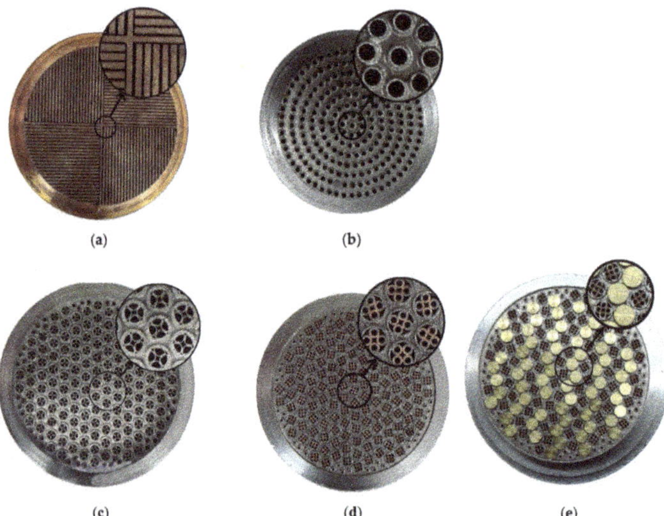

Figure 3. Photographs of (**a**) WCHX 1; (**b**) WCHX 2; (**c**) WCHX 3; (**d**) WCHX 4; and (**e**) WCHX 7. WCHXs 5 and 6 had the same gas channels as WCHX 4.

Table 2. Parameters of the WHCXs.

WHCX	Porosity	Gas-solid Heat Transfer Area	Hydraulic Diameter	Length	Remarks
1	26.2%	0.38 m^2	0.8 mm	64 mm	-
2	23.8%	0.11 m^2	2.44 mm	64 mm	-
3	25.1%	0.261 m^2	1.35 mm	64 mm	-
4	24.5%	0.386 m^2	0.76 mm	64 mm	-
5	24.5%	0.386 m^2	0.76 mm	64 mm	No welding between small and big tubes
6	24.5%	0.193 m^2	0.76 mm	32 mm	-
7	12.3%	0.193 m^2	0.76 mm	64 mm	-

3. Experimental Results

3.1. Influence of Different Configurations

To study the effect of the structure of the water cooler on the heat transfer performance, the performance of WCHX 1, WCHX 2, and WCHX 4, were compared. Figure 4 presents the dependence of the overall relative Carnot efficiency η on the input acoustic power.

$$\eta = \frac{Q(T_h - T_c)}{W_a T_c} \tag{2}$$

where Q is the cooling power, T_c is the cold head temperature, and T_h is the cooling water temperature. A better cooling performance was achieved when WCHX 4 was used. When the input acoustic power was about 1500 W, the system performance was similar for WCHX 1 and WCHX 2; with a larger input acoustic power, the performance with WCHX 2 was better than that with WCHX 1; however, there was still a gap with that observed with WCHX 4. This was apparent in high-power conditions, where WCHX 4 had a great advantage compared with the other two WCHXs. Taking an input acoustic power of 4000 W as an example, the WCHX 4 system achieved a relative Carnot efficiency of 20.6% and cooling power of 319 W, whereas the system with WCHX 1 only achieved an efficiency of 18% and cooling power of 270 W, and the system with WCHX 2 only achieved an efficiency of 18.7% and cooling power of 281 W.

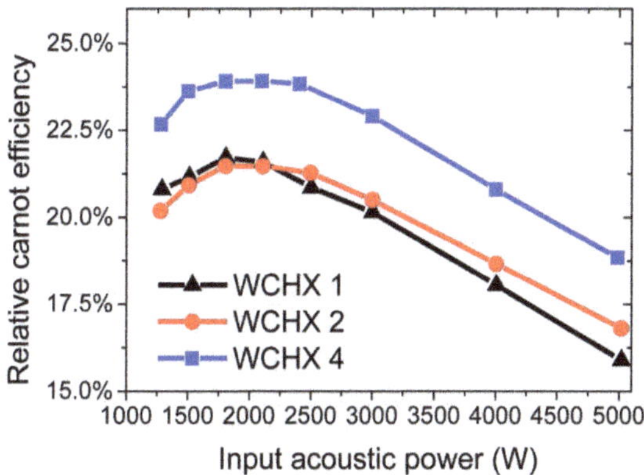

Figure 4. Dependence of relative Carnot efficiency on input acoustic power for WCHXs 1, 2, and 4.

Owing to the aforementioned technical difficulty, the gas temperature was measured at the inlet of the WCHXs, as shown in Figure 5. At 1270 W input acoustic power, there was no large difference in gas temperature among the three WCHX designs: 35 °C with WCHX 1, 24 °C with WCHX 2, and 23 °C with WCHX 4. The gas temperature increased with the input acoustic power, especially in the system with the plated-fin WCHX. When the input acoustic power was 5000 W, the gas temperature was as high as 76 °C with WCHX 1, but only 33 °C with WCHX 4. This implies that WCHX 4 was better cooled.

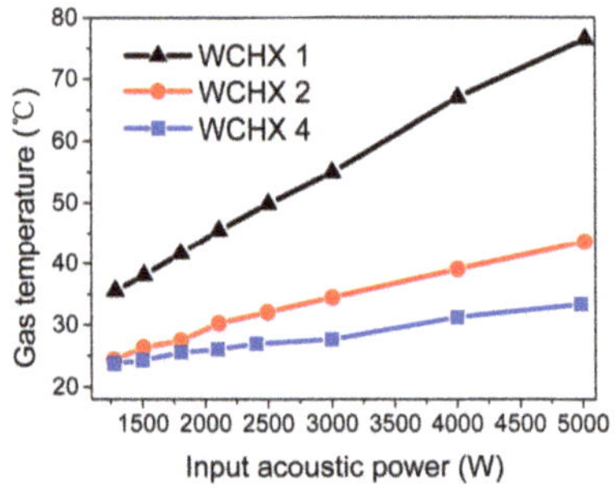

Figure 5. Gas temperatures at the inlets of WCHXs 1, 2, and 4.

What puzzled us is that the gas temperature obtained with WCHX 2 was very close to that with WCHX 4 (Figure 5), but its efficiency was closer to that of WCHX 1 (Figure 4). The temperature distribution along the regenerator shown in Figure 6 may provide some explanations. The x-axis begins at the warm end of the regenerator and ends at its cold end. The data points are the temperatures measured by thermometers T_1 to T_4, and the straight lines are their fitted results. At an input acoustic power of 1270 W, the temperature at the warm end for WCHXs 1 and 2 were very close, and were 10 K higher than that for WCHX 4. This means that with WCHX 4, the temperature difference between the two ends of the regenerator was less, and thus, more cooling power was obtained. At an input acoustic power of 5000 W, the temperature at the warm end observed with WCHX 1 was about 10 K higher than that with WCHX 2, and about 24 K higher than that with WCHX 4. As a result, more effort was required to pump heat from the cold head to the ambient end of the regenerator. Figures 4–6 indicated that the more effective parameter to evaluate the performance of the WCHXs in a pulse tube refrigerator, is the hot-end temperature of the regenerator, instead of the inlet gas temperature of the WCHXs.

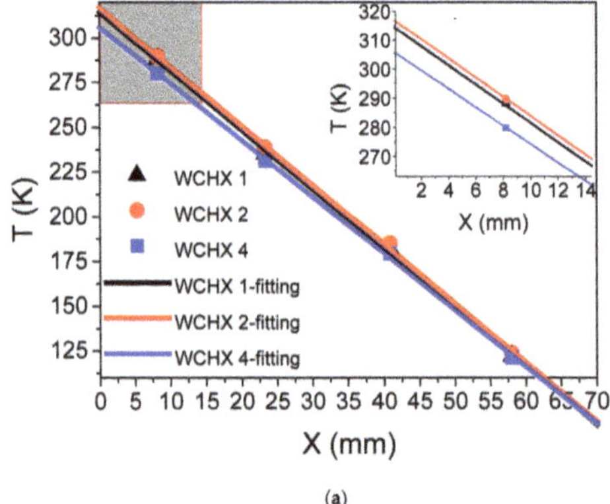

(a)

Figure 6. *Cont.*

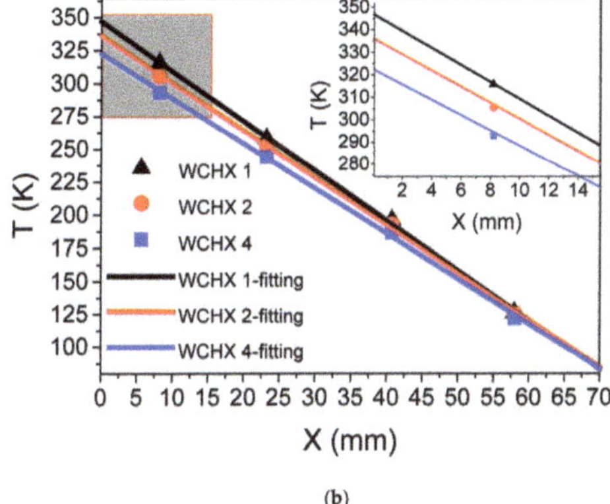

(b)

Figure 6. Temperature distribution in the regenerator at input acoustic power of (**a**) 1270 W and (**b**) 5000 W for WCHXs 1, 2, and 4.

3.2. Influence of Length

Generally, a longer WCHX means a greater heat transfer area, which is helpful for the overall heat transfer, but also results in a greater flow loss. So, an ideal WCHX is a compromise between heat transfer and flow loss. Figure 7 presents the relative Carnot efficiency of a refrigerator with two different length WCHXs. The length of WCHX 6 was half the length of WCHX 4. The system performance of the two WCHX designs was almost the same, because the warm-end temperatures of the regenerator were almost the same, as shown in Figure 8. Thus, the half length WCHX met the heat transfer requirements of the system. The gas temperature at the inlet of the WCHXs was also compared, as shown in Figure 9. The figure shows that the inlet gas temperature of WCHX 4 was smaller than the shorter WCHX 6. This was conducive for reducing the outlet gas temperature of the compressor, which has a positive impact on the stability of the system. From the aspects of improved performance and economy, there is no need to make the WCHX too long.

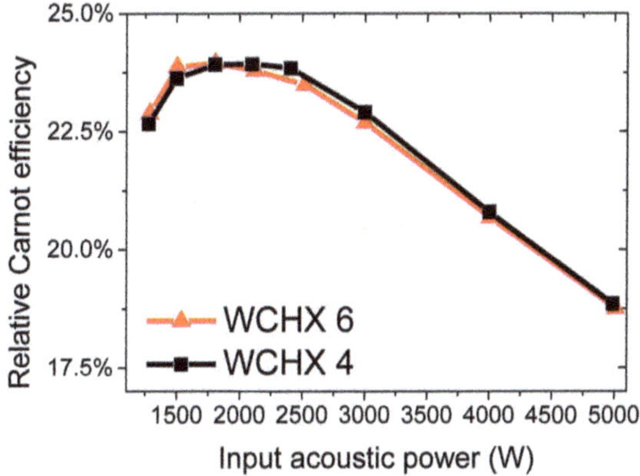

Figure 7. Dependence of relative Carnot efficiency on the input acoustic power for WCHXs 4 and 6.

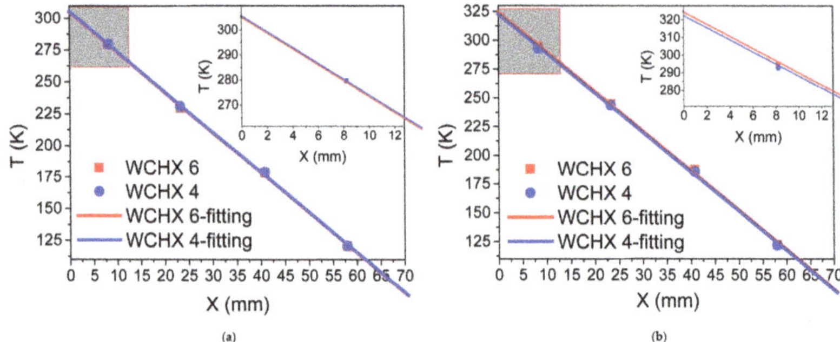

Figure 8. Temperature distribution in the regenerator at an input acoustic power of (**a**) 1270 W and (**b**) 5000 W, for WCHXs 4 and 6.

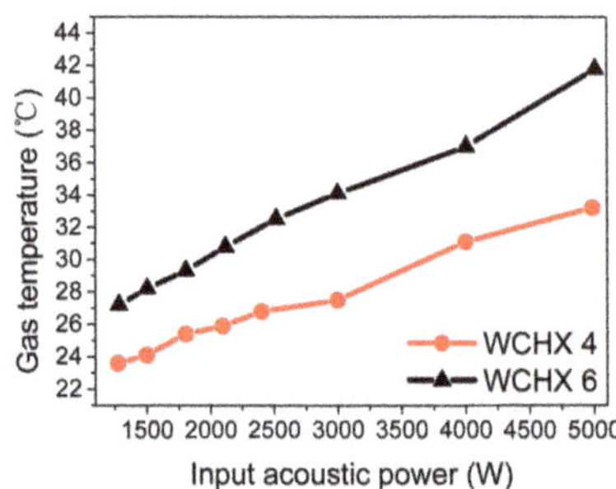

Figure 9. Gas temperature at the inlets of the WCHXs 4 and 6.

3.3. *Influence of Different Contact Thermal Resistance WCHXs*

In WCHX 4, the small tubes acted as fins and transferred heat to the wall of the large tubes, so the heat transfer between the small and large tubes had a significant impact. The small tubes should be welded to the larger tubes, to ensure the best thermal contact, but welding makes the machining process quite complex. If good thermal contact can be achieved by extrusion, the welding process can be omitted. Figure 10 presents the relative Carnot efficiency of the refrigerator with WCHXs 4 and 5. In WCHX 5, the small tubes were not welded to the large tubes. The performance of the two WCHXs was almost the same. This indicates that good thermal contact between the large and small tubes can be achieved by extrusion.

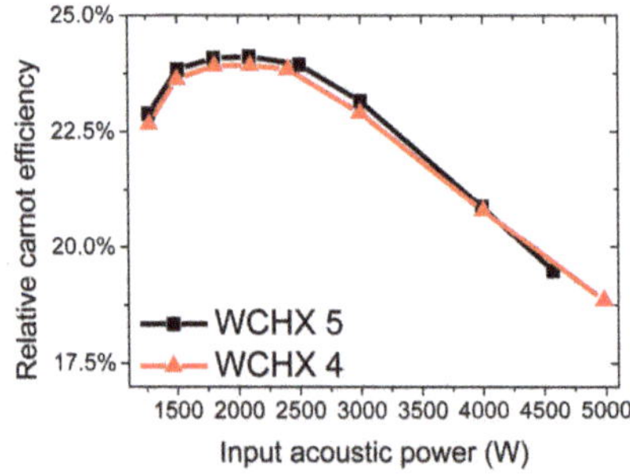

Figure 10. Dependence of relative Carnot efficiency on input acoustic power for WCHXs 4 and 5.

3.4. Influence of Different Hydraulic Diameter

From the definition of the heat transfer coefficient, we know that a smaller hydraulic diameter means a greater heat transfer coefficient, which is conducive to heat transfer. However, the flow resistance between the gas and the wall will be increased if the hydraulic diameter is too small, which will increase the loss in acoustic power. Therefore, the choice of an appropriate hydraulic diameter of the WCHX is very important. Figure 11 presents the relative Carnot efficiency of the refrigerator, with different hydraulic diameters (WCHXs 3 and 4). At the same acoustic power, a better cooling performance was achieved with WCHX 4, which had a smaller hydraulic diameter than WCHX 3. It should be noted that the actual heat transfer area of WCHX 3 was less than that of 4, and it was difficult to distinguish which was the more important factor for causing this different performance. When compared with WCHX 3, the heat transfer in WCHX 4 was increased by about 50%, and the hydraulic diameter was decreased from 1.35 mm to 0.76 mm, but the efficiency in Figure 11 only increased a little. Therefore, we did not think that it was necessary to further decrease the hydraulic diameter.

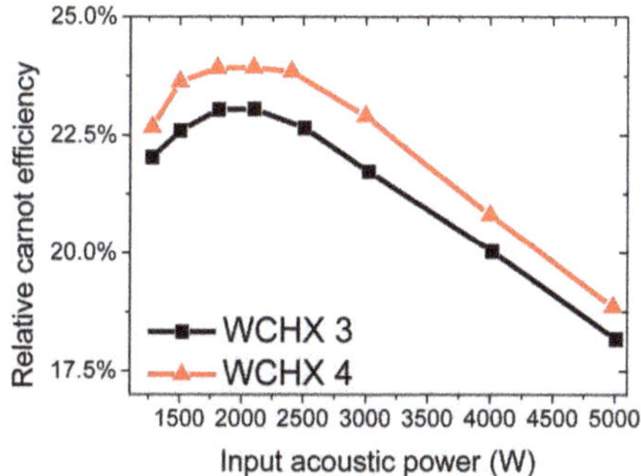

Figure 11. Dependence of relative Carnot efficiency on input acoustic power for WCHXs 3 and 4.

3.5. Influence of Different Porosity

A higher porosity of WCHXs is helpful for decreasing the flow resistance, but will lower the flow velocity and weaken the heat transfer. There were 142 large tubes and 568 small tubes in WCHX 4. There is no doubt that this is a huge number. If the WCHX can meet the heat transfer requirements of the system with a reduced number of tubes, it will save material and produce lower costs during processing. Therefore, we studied the effect of such a reduction on WCHX 4. Half of the gas flow channels were plugged, so that the porosity of the WCHX became 12.3% (WCHX 7). Figure 12 presents the relative Carnot efficiency of the refrigerator with WCHXs 4 and 7. Figure 12 shows that, at the same acoustic power, a better cooling performance was achieved when WCHX 4 was used. Compared with WCHX 7, the porosity and heat transfer area in WCHX 4 was doubled. It can be seen that the efficiency improvement was not significant. The efficiency improvement can be partly explained by the increase of the heat transfer area. Therefore, we loosely argue that the porosity effect is not significant in this experiment. However, it may seriously affect the flow loss and flow inhomogeneity in the regenerator, if the porosity is too small.

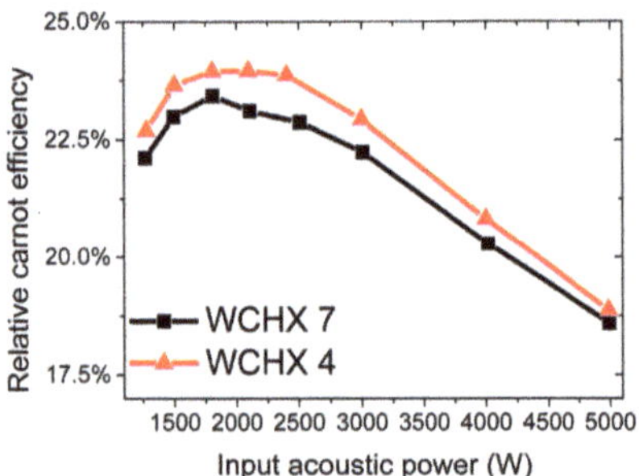

Figure 12. Dependence of relative Carnot efficiency on input acoustic power for WCHXs 4 and 7.

4. Conclusions

Several WCHXs were tested for their use in a high-capacity pulse-tube cooler. The performance of the pulse tube refrigerator with differently structured WCHXs was compared and analyzed. It was found that the warm-end temperature of the regenerator, instead of the gas temperature at the inlet of the WCHX, influenced the cooling performance of the refrigerator. Owing to its more uniform fin temperature distribution, smaller hydrodynamic diameter, and greater heat transfer area, the WCHX design invented by Hu always outperformed the plated-fin and traditional shell-tube type WCHXs. With this WCHX in a pulse tube refrigerator working at a frequency of about 55 Hz, a hydraulic diameter of 0.76 mm, length of 32 mm, and porosity of 24.5%, were appropriate. It was also found that the welding process could be omitted because good thermal contact was achieved by extrusion. This will be helpful for lowering the cost of this configuration. The configuration is suitable for consideration for further use in other regenerative prime movers and coolers.

Acknowledgments: This work was supported by the National Natural Science Foundation of China (Grant No. 51576204) and National Key Research and Development Program of China (Contract No. 2016YFB0901403).

Author Contributions: Jianying Hu, Limin Zhang and Ercang Luo conceived and designed the experiments; Wei Wang and Jingyuan Xu performed the experiments; Wei Wang analyzed the data; Wei Wang and Jianying Hu wrote the paper.

Conflicts of Interest: The authors declare no conflict of interest.

References

1. Gifford, W.E.; Longsworth, R.C. Pulse-tube refrigeration. *J. Eng. Ind.* **1964**, *86*, 264–268. [CrossRef]
2. Zia, J.H. A commercial pulse tube cryocooler with 200 W refrigeration at 80 K. In *Cryocoolers 13*; Ross, R.G., Ed.; Springer: Boston, MA, USA, 2005; pp. 165–171.
3. Zia, J.H. A pulse tube cryocooler with 300 W refrigeration at 80 K and an operating efficiency of 19% carnot. *Cryocoolers* **2007**, *14*, 141–147.
4. Hu, J.Y.; Zhang, L.M.; Zhu, J.; Chen, S.; Luo, E.C.; Dai, W.; Li, H.B. A high-efficiency coaxial pulse tube cryocooler with 500 W cooling capacity at 80 K. *Cryogenics* **2014**, *62*, 7–10. [CrossRef]
5. Potratz, S.A.; Nellis, G.F.; Maddocks, J.R.; Kashani, A.; Helvensteijn, B.P.M.; Rhoads, G.L.; Flake, B. Development of a large-capacity, stirling-type, pulse-tube refrigerator. *AIP Conf. Proc.* **2006**, *823*, 3–10.
6. Potratz, S.A.; Abbott, T.D.; Johnson, M.C.; Albaugh, K.B. Stirling-type pulse tube cryocooler with 1 kW of refrigeration at 77 K. *AIP Conf. Proc.* **2008**, *985*, 42–48.
7. Hu, J.Y.; Luo, E.; Wu, Z.; Yu, G.; Dai, W. Design of a large-capacity multi-piston pulse tube cryocooler. *AIP Conf. Proc.* **2012**, *1434*, 540–546.
8. Hu, J.Y.; Chen, S.; Zhu, J.; Zhang, L.M.; Luo, E.C.; Dai, W.; Li, H.B. An efficient pulse tube cryocooler for boil-off gas reliquefaction in liquid natural gas tanks. *Appl. Energy* **2015**, *164*, 1012–1018. [CrossRef]

9. Caughley, A.; Emery, N.; Nation, M.; Allpress, N.; Kimber, A.; Branje, P.; Reynolds, H.; Boyle, C.; Meier, J.; Tanchon, J. Commercial pulse tube cryocoolers producing 330 W and 1000 W at 77 K for liquefaction. *IEEE Trans. Appl. Supercond.* **2016**, *26*, 1–4. [CrossRef]
10. Bretagne, E.; François, M.X.; Ishikawa, H. Investigations of acoustics and heat transfer characteristics of thermoacoustic driven pulse tube refrigerators. *AIP Conf. Proc.* **2004**, *710*, 1687–1695.
11. Ki, T.; Jeong, S. Optimal design of the pulse tube refrigerator with slit-type heat exchangers. *Cryogenics* **2010**, *50*, 608–614. [CrossRef]
12. Yang, K.X.; Wu, Y.N.; Zhang, A.K.; Xiong, C. Fabrication of Taper Gap Warm End Heat Exchanger in Coaxial Pulse Tube Cryocooler. *Adv. Mater. Res.* **2012**, *591–593*, 365–368. [CrossRef]
13. Chen, Y.Y.; Luo, E.C.; Dai, W. Heat transfer characteristics of oscillating flow regenerator filled with circular tubes or parallel plates. *Cryogenics* **2007**, *47*, 40–48. [CrossRef]
14. Gholamrezaei, M.; Ghorbanian, K. Thermal analysis of shell-and-tube thermoacoustic heat exchangers. *Entropy* **2016**, *18*, 301. [CrossRef]
15. Kamsanam, W.; Mao, X.; Jaworski, A.J. Thermal performance of finned-tube thermoacoustic heat exchangers in oscillatory flow conditions. *Int. J. Therm. Sci.* **2016**, *101*, 169–180. [CrossRef]
16. Jaworski, A.J.; Piccolo, A. Heat transfer processes in parallel-plate heat exchangers of thermoacoustic devices–numerical and experimental approaches. *Appl. Therm. Eng.* **2012**, *42*, 145–153. [CrossRef]
17. Tang, K.; Yu, J.; Jin, T.; Gan, Z.H. Influence of compression-expansion effect on oscillating-flow heat transfer in a finned heat exchanger. *J. Zhejiang Univ. Sci. A* **2013**, *14*, 427–434. [CrossRef]
18. Tang, K.; Yu, J.; Jin, T.; Wang, Y.P.; Tang, W.T.; Gan, Z.H. Heat transfer of laminar oscillating flow in finned heat exchanger of pulse tube refrigerator. *Int. J. Heat Mass Transfer* **2014**, *70*, 811–818. [CrossRef]
19. Xu, J.Y.; Hu, J.Y.; Zhang, L.M.; Luo, E. A novel shell-tube water-cooled heat exchanger for high-capacity pulse-tube coolers. *Appl. Therm. Eng.* **2016**, *106*, 399–404. [CrossRef]
20. Luo, L.A.; D'Ortona, U.; Tondeur, D. Compact heat exchangers. In *Microreaction Technology: Industrial Prospects: Imret 3: Proceedings of the Third International Conference on Microreaction Technology*; Ehrfeld, W., Ed.; Springer: Berlin/Heidelberg, Germany, 2000; pp. 556–565.
21. Kuppan, T. *Heat Exchanger Design Handbook*, 2nd ed.; CRC Press: Boca Raton, FL, USA, 2013; p. 21.
22. Hu, J.Y.; Long, X.D.; Li, H.B. Heat exchanger, oscillation flow system and heat exchanger processing methods, CN104964585A [P/OL]. 2015.
23. Bouvier, P.; Stouffs, P.; Bardon, J.-P. Experimental study of heat transfer in oscillating flow. *Int. J. Heat Mass Transfer* **2005**, *48*, 2473–2482. [CrossRef]
24. Chen, Y.; Luo, E.; Dai, W. Heat transfer characteristics of oscillating flow regenerators in cryogenic temperature range below 20 K. *Cryogenics* **2009**, *49*, 313–319. [CrossRef]
25. Hu, J.Y.; Wang, W.; Luo, E.C.; Chen, Y.Y. Acoustic field modulation in regenerators. *Cryogenics* **2016**, *80*, 1–7. [CrossRef]

© 2017 by the authors. Licensee MDPI, Basel, Switzerland. This article is an open access article distributed under the terms and conditions of the Creative Commons Attribution (CC BY) license (http://creativecommons.org/licenses/by/4.0/).

Article

Comparative Performance of Thermoacoustic Heat Exchangers with Different Pore Geometries in Oscillatory Flow. Implementation of Experimental Techniques

Antonio Piccolo [1,*], Roberto Siclari [1], Fabrizio Rando [1] and Mauro Cannistraro [2]

[1] Department of Engineering, University of Messina, Contrada di Dio, 98166 S. Agata (Messina), Italy; robsiclari@gmail.com (R.S.); fabrizio.rando@studenti.unime.it (F.R.)
[2] Department of Architecture, University of Ferrara, Via della Ghiara 36, 44121 Ferrara, Italy; mauro.cannistraro@unife.it
* Correspondence: apiccolo@unime.it; Tel.: +39-090-397-7311

Received: 24 July 2017; Accepted: 31 July 2017; Published: 2 August 2017

Abstract: Heat exchangers (HXs) constitute key components of thermoacoustic devices and play an important role in determining the overall engine performance. In oscillatory flow conditions, however, standard heat transfer correlations for steady flows cannot be directly applied to thermoacoustic HXs, for which reliable and univocal design criteria are still lacking. This work is concerned with the initial stage of a research aimed at studying the thermal performance of thermoacoustic HXs. The paper reports a detailed discussion of the design and fabrication of the experimental set-up, measurement methodology and test-HXs characterized by two different pore geometries, namely a circular pore geometry and a rectangular (i.e., straight fins) pore geometry. The test rig is constituted by a standing wave engine where the test HXs play the role of ambient HXs. The experiment is conceived to allow the variation of a range of testing conditions such as drive ratio, operation frequency, acoustic particle velocity, etc. The procedure for estimating the gas side heat transfer coefficient for the two involved geometries is described. Some preliminary experimental results concerning the HX with straight fins are also shown. The present research could help in achieving a deeper understanding of the heat transfer processes affecting HXs under oscillating flow regime and in developing design optimization procedures.

Keywords: thermoacoustics; heat exchangers; heat transfer; acoustic power

---

1. Introduction

This paper addresses the issues of heat transfer in heat exchangers working under oscillatory flow conditions, which are typically found in thermoacoustic devices. Thermoacoustic engines are a new class of energy conversion devices (prime movers, refrigerators and heat pumps) whose operation relies on the interaction between heat and sound in close proximity of solid surfaces, a phenomenon identified as "thermoacoustic effect" [1,2]. Since in these devices the synchronization among the compression, expansion and heat transfer phases of the thermodynamic gas cycle is naturally accomplished by an acoustic wave, a number of technological benefits directly result. First of all, there is the complete absence of moving mechanical parts (pistons, sliding seals, etc.) which leads to engineering simplicity, reliability, longevity and low maintenance costs. Secondly, they are intrinsically low cost devices, being constituted basically by a small number of standard components made of inexpensive and common materials. Furthermore, they use environment friendly working fluids, can be employed in a large variety of applications (those involving heating, cooling or power generation [3–5]) and can be driven by different power sources (gas/biomass combustion, solar energy, waste heat, etc.).

These characteristics make thermoacoustic technology a discipline of relevant interest for the energy industry giving it a primary position among emerging renewable energy technologies.

A typical thermoacoustic device consists of (a) an acoustic network (acoustic resonator), (b) an electro-acoustic transducer, (c) a porous solid medium (namely a regenerator in travelling-wave systems [6] or a stack in standing-wave systems [7]) and (d) at least a pair of heat exchangers (HXs) [8]. The stack/regenerator is the component where the desired heat/sound energy conversion takes place. "Hot" and "Cold" HXs, placed in close proximity of both ends of the stack/regenerator, absorb or supply heat from its ends thus enabling heat communication with external heat sources and sinks.

Heat exchangers constitute, in addition to the stack/regenerator, the highly dissipative components of thermoacoustic engines. Their porous structure entails, in fact, considerable flow resistance while steep thermal gradients are generally imposed on them to sustain the required heat fluxes. When designing efficient HXs aimed at transferring a target heat load the HX length along the direction of acoustic oscillation and the pore hydraulic radius should be simultaneously optimized in order to:

- Provide the heat transfer surface area compatible with minimum acoustic power loss caused by thermal and viscous dissipation;
- Provide the temperature drop between the HX and the adjacent fluid compatible with minimum thermal irreversibility associated to heat transfer [9,10]. An optimized HX should be able to achieve high transfer rates under small temperature differences in conjunction to low acoustic dissipation.

Although thermoacoustic HXs constitute fundamental components of thermoacoustic devices their design constitute up to now a technical and engineering challenge. This is due to the fact that the flow is oscillatory and the existing knowledge for steady flow arrangements is of little practical value [11–13]. As a consequence, standard heat transfer correlations for steady flows cannot be directly applied to thermoacoustic HXs, for which reliable and unambiguous design criteria are still lacking.

This work is concerned with the initial stage of a research aimed at studying the thermal performance of thermoacoustic HXs working under oscillatory flow conditions. The paper reports a detailed discussion of the design and fabrication of the experimental set-up, measurement methodology and test-HXs characterized by two different pore geometries, namely a circular pore geometry and a rectangular (i.e., straight fins) geometry. Some preliminary experimental results concerning the HX with straight fins are also shown. The research is aimed at achieving a deeper insight into the heat transfer processes affecting HXs under oscillating flow regime. This could help in developing optimization procedures of their performance in the design phase.

2. The Experimental Set-Up

The test rig considered for this research is a home-made standing wave thermoacoustic engine using air at atmospheric pressure as working fluid. A schematic of the engine is shown in Figure 1. The engine comprises four basic elements, namely an acoustic resonator filled with air at atmospheric pressure, a stack, an ambient heat exchanger and an electric heater. The engine works by heating one edge of the stack by the heater while anchoring the other edge to ambient temperature by the ambient HX. Acoustic waves will spontaneously generate in the device when the temperature gradient across the stack overcomes an onset level (depending mainly on the length and position of the stack inside the resonator).

- The resonator is made of stainless steel and has a variable cross section geometry to suppress harmonic generation [14] and to match the target operation frequency (155 Hz) by a shorter total length compared to a straight (constant diameter) resonator. It is essentially a pressure vessel defining the working frequency of the system and the pressure-velocity phase relationship of the acoustic wave. The resonator comprises a "hot" duct with a diameter of 10.5 cm and 10 cm long, a stack holder with a diameter of 10.5 cm and 8 cm long and a narrow duct with a diameter of 5.3 cm and 20 cm long. The wide ducts ("hot" duct and stack holder) and the narrow duct are coupled by a conical shaped section and terminate in a buffer volume. The resonance frequency of the rig can be varied by modifying its length through additional segments.

- The stack is a honeycomb stack made of a ceramic material (Celcor) characterized by a low thermal conductivity and a high specific heat. These properties enhance the stack performance in converting heat to acoustic power [15]. It fits the stack holder and is 8 cm long in the direction of the resonator axis (the longitudinal direction of acoustic vibration). The pores have a squared shape and are characterized by a hydraulic radius of around 0.28 mm. The cell density is near 400 cpsi. The stack placed inside the stack-holder and a magnified view of a portion of it are shown in Figure 2.
- The heater is fabricated by passing about 4 m of a Ni-Cr wire (30 Ω/m) uniformly through the pores of a honeycomb ceramic slice 1 cm long along the resonator axis (and made of the same material of the stack) that served as a support. The slice is inserted on one side of a stainless steel ring, 2 cm long along the resonator axis, equipped with feedthroughs for power cables. The heater is placed between the "hot" duct and the stack holder. It is fed through a variable voltage autotransformer and is able to dissipate up to 600 W of electric power. A picture of the heater is shown in Figure 3.
- The HXs under test (widely descripted in the next section) function in this arrangement as ambient HXs (i.e., subtract heat from the gas oscillating at the ambient side of the stack) and are placed between the stack holder and the conical shape section at a distance of about 20 cm from the closed end of the "hot" duct. This location is around midway the velocity node and antinode and allows a wide range of velocity amplitudes (or, equivalently, gas displacement amplitudes) to be selected by varying the heat input to the system. This information has been deduced by simulating the engine behavior by highly specialized design tools of thermoacoustics systems, namely the DelatEC computer code (Design Environment for Low-Amplitude ThermoAcoustic Engines) [16], developed at the Los Alamos National Laboratory (see Section 5).

Note that the test-rig arrangement above described do not correspond to a requirement of performance optimization but, rather, to the need of guaranteeing that the HXs under test (1) be interested by substantial heat transfer rates (and associated temperature changes in the cooling water) and (2) by values of the local acoustic particle displacement ranging around (both below and above) the axial length of the HXs, as it will be more exhaustively explained in Section 5.

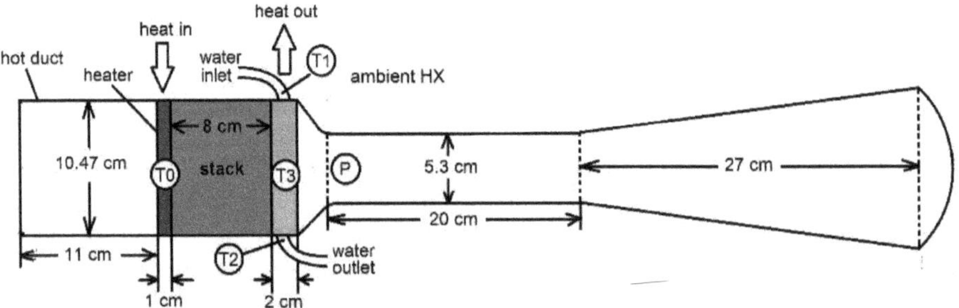

Figure 1. Schematic of the standing wave test engine. Labels "T" indicate thermocouple locations. Label "P" indicates microphone location.

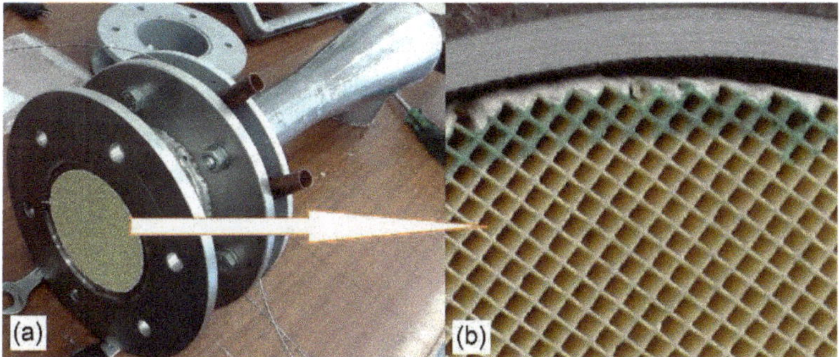

Figure 2. (**a**)The stack placed inside the stack-holder; (**b**) a magnified view of a portion of the stack.

Figure 3. The heater.

3. The Heat Exchanger Prototypes

The HXs prototypes investigated in this study have a similar structure but differentiate each other essentially for the pore geometry. These two home-made water-cooled HXs, in fact, are both made of copper, have the same length along the resonator axis (2 cm), have pores of almost equal hydraulic radius and have the same flow arrangement (cross flow with four water-tube passes). The comparative test of these devices should therefore allow to study the impact of the pore geometry on the performance of the HXs and to gain information on the related gas-side heat transfer coefficient.

The first ambient HX (AHX1) has a rectangular pore geometry being of the fin-and-tube type. For its fabrication a copper ring 2 cm thick with an internal radius of 10.5 cm and an external radius of 14 cm has been considered as mechanical support of the HX itself. Just outside the HX the cooling water flows through a copper tube with a diameter of 14 mm which has been crushed to obtain a rectangular cross section pipe of area 2 cm $\times$ 0.5 cm. The crushed pipe has then been bent six times to allow for four passages in the front section of the HX as shown in Figure 4a. The bent segments don't obstruct the open area of the HX since they have been inserted in compartments extracted from the ring support. The fins are made of laminated copper 0.35 mm thick and are spaced 1 mm apart. They have been welded to the external walls of the pipe by a strong brazing with phosphor-copper rods. The blockage ratio (*BR*) of this HX, i.e., the ratio of the open to the total cross section area, is calculated to be around 55%. The hydraulic radius of the pores is R_h = 0.5 mm.

The second ambient heat exchanger (AXH2) is fabricated from a circular copper block. Gas passages are obtained by drilling 574 holes with a diameter of 2.5 mm in parallel to its axis. Cooling water passages are obtained by drilling 4 channels with a diameter of 5 mm perpendicularly to the

HX axis, as shown in Figure 4b. The water also flows partially around the perimeter of the block. The blockage ratio of this HX is calculated to be around 33%. The hydraulic radius of the pores is $R_h \approx 0.6$ mm. Compared to AHX1 this heat exchanger is characterized by a simpler fabrication technique but the circular pore geometry hardly allows to get blockage ratios exceeding 40%.

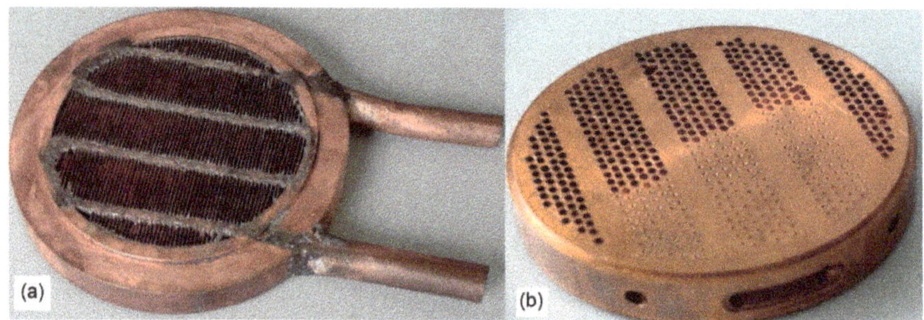

Figure 4. (**a**) The ambient heat exchanger with rectangular pore geometry; (**b**) The ambient heat exchanger with circular pore geometry.

Both HXs are characterized, as above specified, by pores of almost equal hydraulic radius (R_h = 0.5 mm and 0.6 mm). This is a precise design choice since, as inferred by simulations, at the design working frequency of the engine ($f \approx 155$ Hz) the thermal penetration depth of the gas

$$\delta_\kappa = \sqrt{\frac{2\kappa}{\omega}}, \tag{1}$$

i.e., the distance that heat can diffuse in an acoustic cycle, amount approximately to $\delta_\kappa \approx 0.22$ mm so that $R_h/\delta_\kappa \approx 2$. This assures a good thermal contact between the oscillating gas and the HX solid walls. In Equation (1) κ and ω are the gas thermal diffusivity and the angular frequency of the acoustic wave respectively.

A picture of the standing wave engine in place for testing without and with thermal insulation (rock wool) is shown in Figure 5a,b respectively.

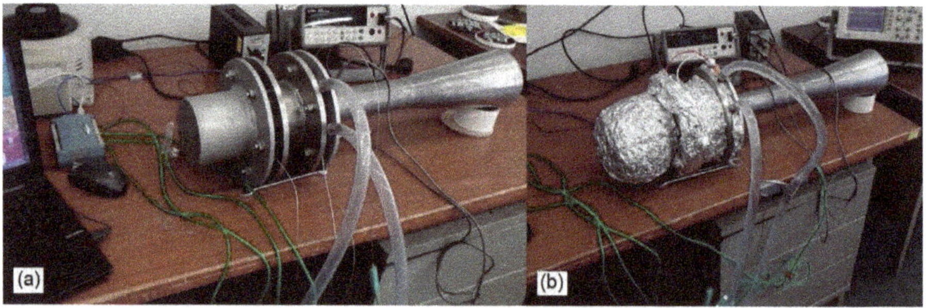

Figure 5. The standing wave test engine; (**a**) without thermal insulation (rock wool) and (**b**) with thermal insulation.

4. The Measurement Methodology

To get information of the heat transfer rates sustained by the HXs and on the related gas-side heat transfer coefficient experimental and calculation procedures based on standard energy balance measurement are developed. As for the first parameter, the methodology involves the determination of the temperature at the inlet section of the water flow (T_{wi}), the one at the outlet section (T_{wo}) and the mass flow rate of water ($\dot{m}_w$). This allows to calculate the thermal power captured by the HX as

$$\dot{Q}_A = \dot{m}_w c_w (T_{wo} - T_{wi}), \tag{2}$$

where c_w is the water specific heat. Note that the standing wave test rig discussed so far is not connected to any acoustic load. In other words, there is no component (loudspeakers, linear alternator, resistance/compliance series assembly, etc.) which could extract a fraction of the acoustic power produced by the stack from the engine. More exactly, the acoustic load is constituted by the internal surfaces of the resonator and by the HXs solid walls themselves over which the acoustic power is entirely consumed by viscous dissipation and thermal relaxation processes. Therefore, in the absence of thermal and/or acoustic power leakages to/from the outside environment, the heat extracted by the ambient HX, and represented by Equation (2), should precisely coincide with the heat delivered by the heater to the system ($\dot{Q}_H$). To reduce uncertainty in measurements the system (from the "hot" duct to the AHX) is thermally insulated from the external ambient by rock wool.

To estimate the gas-side heat transfer coefficient a number of thermocouples need to be placed/embedded in the HXs with a radial distribution to measure the temporally and spatially averaged temperatures of the solid walls and of the gas locally oscillating. A mean heat transfer coefficient can be then calculated from the ratio of the heat transfer rate per unit exchange area and the temperature difference between the gas and the solid walls of the HX. For AHX1 a different procedure, involving the measurement of the temporally and spatially averaged temperature of the gas in the HX (T_g) near the ambient side of the stack can be applied [11]. For a cross-flow heat exchanger, in fact, the heat transfer rate can be expressed as

$$\dot{Q}_A = U'A\theta_{ml}, \tag{3}$$

where $U'A$ is the overall thermal conductance of the HX and θ_{ml}, the log-mean temperature difference, is defined as

$$\theta_{ml} = F\frac{T_{wo} - T_{wi}}{\ln\left(\frac{T_g - T_{wi}}{T_g - T_{wo}}\right)}, \tag{4}$$

F being a correction factor for cross flow HXs and having taken into account that, since the flow is oscillatory, the inlet and outlet temperatures of the primary fluid (the gas) are the same (T_g). Once calculated the values of $\dot{Q}_A$ and θ_{ml} from the experimental measurements it is possible to evaluate the overall thermal conductance of the HX from Equation (3). To estimate the gas-side heat transfer coefficient the overall conductance $U'A$ must now be expressed in an explicit form. A simplified heat transfer model provides [17].

$$U'A = \frac{1}{\frac{1}{h_w A_i} + \frac{s}{K_t A_i} + \frac{1}{h\left[\sum_i (\prod L_{fi})\eta_{fi} + A_b\right]}}, \tag{5}$$

where h_w is the water-side heat transfer coefficient, A_i is the total inner surface of the tubes, K_t is the thermal conductivity of the tube material (copper), s is the wall thickness of the tube, h is the gas-side heat transfer coefficient, $\prod$ is the perimeter of the fin cross section, L_{fi} is half the length of the i-th fin, A_b is the total area of the unfinned external surface of the tubes and where η_{fi}, the efficiency of the i-th fin, is defined as

$$\eta_{fi} = \frac{\tanh(mL_{fi})}{mL_{fi}}, \quad m = \sqrt{\frac{h\prod}{A_f K_f}} \tag{6}$$

A_f and K_f being the cross section area and thermal conductivity of the fin respectively. In deducing Equation (5) it has been considered that the tubes for the passage of the water have a rectangular cross section so that the associated thermal resistance can be approximated by that of a plane wall.

The heat transfer coefficients derived by means of the above method have to be correlated to the local oscillating velocity or, more exactly, to the acoustic Reynolds number $\mathrm{Re}_1 = \rho_0 D_h v/\eta$ (ρ_0 being the mean gas density, D_h the hydraulic diameter of the HX pore, v the acoustic velocity amplitude and η the dynamic viscosity). This last is derived by modelling the standing-wave engine by the DeltaEC code and evaluating the acoustic particle velocity at the position of the ambient HX. The validation of the data involves an iterative refinement of the model by the acoustic pressure values measured in different positions along the resonator axis by dynamic pressure sensors.

5. Simulations with DeltaEC

To model the thermoacoustic engine and get information on the acoustic field neat the test HXs the computer code DeltaEC has been used. The code integrates the linear theory of thermoacoustics, firstly formulated by Rott [18] and subsequently refined by Swift [1,2], which has demonstrated accurate precision for pressure amplitudes up to 10% of the mean pressure. The code performs 1D numerical integration of the momentum, continuity and energy equations through each segment of the acoustic network. The solutions found for adjoining segments are then matched by imposing the continuity of pressure and volume flow rate at their junction. For further details the reader is addressed to ref. [16].

In Figure 6 the axial distribution of the acoustic field (acoustic pressure, p, and volumetric velocity, U) is shown. The p and U profiles resemble those characterizing a half-wavelength resonator. The graph shows how the HX location falls between a velocity node and antinode. The acoustic particle velocity near the HX can be modulated by varying the heat input to the system. This also influences the operative temperature of the heater, T_H, the acoustic pressure amplitude, p, and the amplitude of the particle oscillation displacement

$$x_1 = \frac{U}{A_{HX}\omega} \tag{7}$$

A_{HX} being the cross section area of the HX open to gas flow. The last is an important parameter in the design of HXs since, being the flow oscillatory, an excessively long HX (along the longitudinal direction of the particle acoustic oscillation) could only lead to additional thermoviscous losses without incrementing the surface area available for heat transfer.

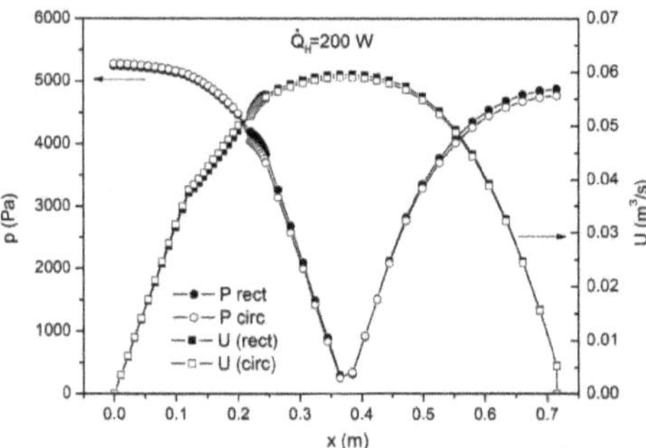

Figure 6. The axial distribution of the acoustic pressure amplitude and volumetric velocity amplitude.

Conversely, a much too short HX could be ineffective in transferring the required thermal load. It is generally assumed, as a rule of tumb, that the optimal length of the HX (L_{HX}) should be of the order of the peak-to-peak acoustic displacement amplitude [1], although this problem remains to date an open issue.

In Figure 7 the variables T_H, DR and $2x_1/L_{HX}$ are reported as a function of the heat input for the two typologies of HX. In both cases the drive ratio is lower than 10% even at the highest power delivered by the heater so the growth of non linear acoustic effects should not considerably affect the engine operation. But, more interesting, is the circumstance that at a given $\dot{Q}_H$ value the peak-to-peak particle displacement amplitude, $2x_1$, surpasses the length of the HX. This "transition" should allow to investigate on the optimal length along the longitudinal direction which maximizes the HX performance.

As a concluding remark it has to be underlined that DeltaEC relies, as for the calculation of the gas side heat transfer coefficient, on a simple boundary layer model expressed by the equation

$$h = \frac{K}{y_{eff}}, \tag{8}$$

where

$$y_{eff} = \min\{\delta_K, R_h\}, \tag{9}$$

It is retained that this expression may be quite inaccurate and be affected by an error of about a factor of two.

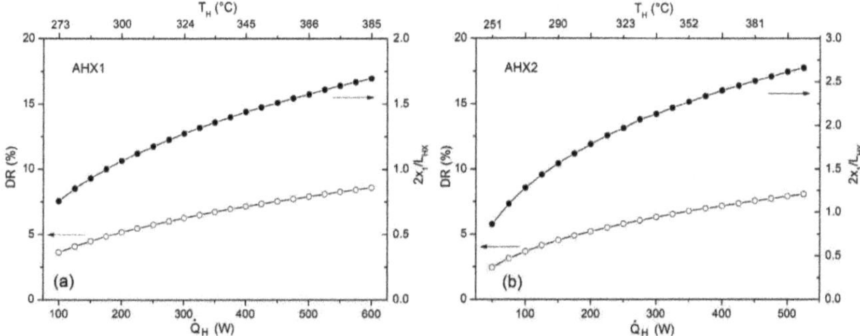

Figure 7. The drive ratio and peak-to-peak particle displacement amplitude as a function of the input heat flux for (**a**) the HX with rectangular pore geometry and (**b**) the HX with circular pore geometry. The operation temperatures of the heater, T_H, are also indicate on the horizontal scale on the top.

6. Preliminary Experimental Results

In Figures 8 and 9 some preliminary experimental results concerning AHX1 are shown. In the first graph the difference between the temperatures of the hot HX (the heater) and of the ambient HX is reported as a function of the measured input heat transfer rate ($\dot{Q}_H$). The thermocouple (K-type thermocouple) of the heater was placed inside a pore of the ceramic honeycomb support so it presumably measured a mean value (T_H) between the gas and the solid wall temperatures.

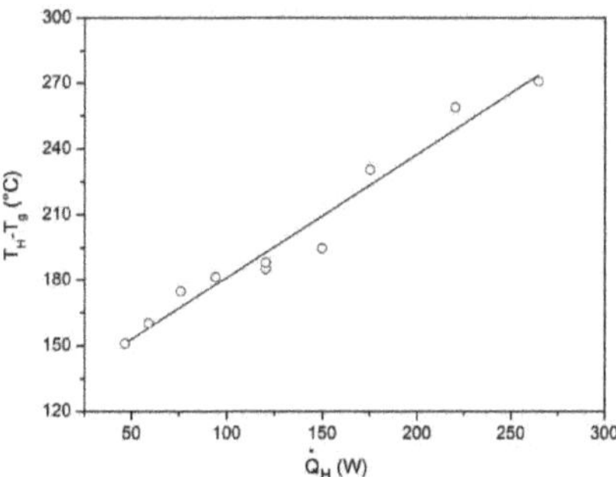

Figure 8. The temperature difference between the hot and ambient heat exchanger with rectangular pores as a function of the input heat.

The thermocouple of the ambient HX was suspended by a little spacer in the gap between two fins so in this case the gas temperature (T_g) was measured. The input heat flux $\dot{Q}_H$ was calculated by measuring the voltage drop across the heater and the flowing current. The heat flux extracted by the ambient HX, $\dot{Q}_A$, was calculated through Equation (2) from the measured values of water mass flow rate ($\dot{m}_w \approx 0.052$ kg/s) and of its temperature at the inlet ($T_{wi} \approx 300$ K) and outlet sections of the HX measured by two thermocouples. The acoustic pressure values reported in Figure 9 were measured by a condenser microphone (sensitivity 1.589 mV/Pa) flush mounted in the wall of the narrow duct at a distance of 3 cm from the ambient HX (see Figure 1).

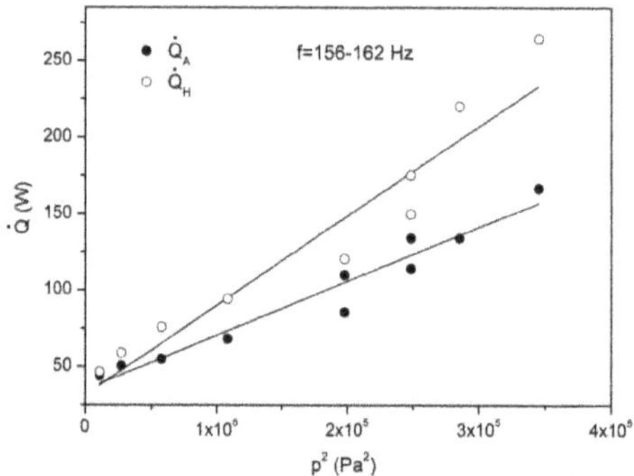

Figure 9. The heat introduced by the heater and extracted by the ambient heat exchanger with rectangular pores as a function of the square of the acoustic pressure.

The air starts to oscillate spontaneously in the engine for $T_H \approx 200$ °C at a (resonance) frequency of 156.5 Hz (which slightly grew with $\dot{Q}_H$). Figure 8 shows how the temperature difference (T_H-T_g) grows quickly and almost linearly with $\dot{Q}_H$ in the investigated range. A linear behavior is also found in Figure 9 where $\dot{Q}_H$ and $\dot{Q}_A$ are reported as a function of the square of the acoustic pressure. Note, however, as the difference between $\dot{Q}_H$ and $\dot{Q}_A$ (increasing with the heater power) is a clear evidence of heat and/or acoustic power leakage toward the external environment, so an improvement of the engine insulation is required.

These findings are in accordance with the linear theory of thermoacoustics and let suppose that in the investigated operation range non linear effects are negligible. This result is somehow expected when considering that at the highest $\dot{Q}_H$ values shown in Figure 9 the measured acoustic pressure reached 0.6% of the mean pressure. Since the microphone is located around halfway a pressure node and antinode this value should roughly correspond to a drive ratio of about 1.2%, that is well below the onset level of non linear effects.

7. Conclusions

The research described in the present paper focuses on the development of experimental techniques for investigating on the fundamental issue of heat transfer of thermoacoustic heat exchangers under acoustically oscillating flows. The test rig, a standing wave engine working with air at atmospheric pressure, and the test heat exchangers, a circular pore and rectangular pore cross flow heat exchanger, have been designed and manufactured. The prototype has been conceived to study the performance of the heat exchangers under a range of operating conditions (drive ratio, operation frequency, acoustic particle velocity temperature differences, etc.) and to estimate, by specific procedures, the magnitude of gas side heat transfer coefficient relatively to the two considered pore geometries. The present research, the initial stage of which is discussed in this paper, could help in achieving a deeper understanding of the heat transfer processes affecting heat exchangers under oscillating flow regime and in developing optimization procedures of their performance in the design phase.

Acknowledgments: This research did not receive any specific grant from funding agencies in the public, commercial, or not-for-profit sectors. The authors are grateful to the reviewers for their suggestions which largely contributed to improve the quality of the present paper.

Author Contributions: Antonio Piccolo conceived and designed the experiments, built the heater and the HX with circular pores, performed the experiments and numerical simulations, analysed the data and wrote the paper; Roberto Siclari built the resonator and the HX with parallel fins and contributed to analyse the data; Fabrizio Rando contributed to perform the experiments and analyse the data; Mauro Cannistraro contributed to analyse the data and write the paper.

Conflicts of Interest: The authors declare no conflict of interest.

References

1. Swift, G.W. Thermoacoustic engines. *J. Acoust. Soc. Am.* **1988**, *84*, 1145–1180.
2. Swift, G.W. Thermoacoustics: A unifying perspective for some engines and refrigerators. *J. Acoust. Soc. Am.* **2002**, *113*. [CrossRef]
3. Backhaus, S.; Tward, E.; Petach, M. Traveling-wave thermoacoustic electric generator. *Appl. Phys. Lett.* **2004**, *85*, 1085–1087. [CrossRef]
4. Yu, Z.; Jaworski, A.J.; Backhaus, S. Travelling-wave thermoacoustic electricity generator using an ultra-compliant alternator for utilization of low-grade thermal energy. *Appl. Energy* **2012**, *99*, 135–145. [CrossRef]
5. Piccolo, A. Design issues and performance analysis of a two-stage standing wave thermoacoustic electricity generator. *Sustain. Energy Technol. Assess.* **2016**, in press. [CrossRef]
6. Backaus, S.; Swift, G.W. A thermoacoustic Stirling heat engine: Detailed study. *J. Acoust. Soc. Am.* **2000**, *107*, 3148–3166. [CrossRef]
7. Swift, G.W. Analysis and performance of a large thermoacoustic engine. *J. Acoust. Soc. Am.* **1992**, *92*, 1551–1563. [CrossRef]
8. Garrett, S.L.; Perkins, D.K.; Gopinath, A. Thermoacoustic Refrigerator Heat Exchangers: Design, Analysis and Fabrication. In Proceedings of the Tenth International Heat Transfer Conference, Brighton, UK, 14–18 August 1994; pp. 375–380.
9. Piccolo, A. Numerical study of entropy generation within thermoacoustic heat exchangers with plane fins. *Entropy* **2015**, *17*, 8228–8239. [CrossRef]
10. Piccolo, A. Optimization of thermoacoustic refrigerators using second law analysis. *Appl. Energy* **2013**, *103*, 358–367. [CrossRef]
11. Peak, I.; Braun, J.E.; Mongeau, L. Characterizing heat transfer coefficients for heat exchangers in standing wave thermoacoustic coolers. *J. Acoust. Soc. Am.* **2005**, *118*, 2271–2280. [CrossRef]
12. Jaworski, A.J.; Piccolo, A. Heat transfer processes in parallel-plate heat exchangers of thermoacoustic devices—Numerical and experimental approaches. *Appl. Therm. Eng.* **2012**, *42*, 145–153. [CrossRef]
13. Mohd Saat, F.A.Z.; Jaworski, A.J. The effect of temperature field on low amplitude oscillatory flow within a parallel-plate heat exchanger in a standing wave thermoacoustic system. *Appl. Sci.* **2017**, *7*, 417. [CrossRef]
14. Tijani, M.E.H.; Zeegers, J.C.H.; De Waele, A.T.A.M. Design of thermoacoustic refrigerators. *Cryogenics* **2002**, *42*, 49–57. [CrossRef]
15. Piccolo, A.; Pistone, G. Computation of the time-averaged temperature fields and energy fluxes in a thermally isolated thermoacoustic stack at low acoustic Mach numbers. *Int. J. Therm. Sci.* **2007**, *46*, 235–244. [CrossRef]
16. Clark, J.P.; Ward, W.C.; Swift, G.W. Design environment for low-amplitude thermoacoustic energy conversion (DeltaEC). *J. Acoust. Soc. Am.* **2007**, *122*. [CrossRef]
17. Incoprera, F.P.; De Witt, D.P. *Fundamentals of Heat and Mass Transfer*, 3rd ed.; Wiley and Sons: New York, NY, USA, 1996.
18. Rott, N. Thermoacoustics. *Adv. Appl. Mech.* **1980**, *20*, 35–75.

© 2017 by the authors. Licensee MDPI, Basel, Switzerland. This article is an open access article distributed under the terms and conditions of the Creative Commons Attribution (CC BY) license (http://creativecommons.org/licenses/by/4.0/).

Article

The Effect of Temperature Field on Low Amplitude Oscillatory Flow within a Parallel-Plate Heat Exchanger in a Standing Wave Thermoacoustic System

Fatimah A.Z. Mohd Saat [1] and Artur J. Jaworski [2,*]

[1] Centre for Advanced Research on Energy, Faculty of Mechanical Engineering, Universiti Teknikal Malaysia Melaka, Hang Tuah Jaya, Durian Tunggal 76100, Malaysia; fatimah@utem.edu.my

[2] Faculty of Engineering, University of Leeds, Woodhouse Lane, Leeds LS2 9JT, UK

* Correspondence: a.j.jaworski@leeds.ac.uk; Tel.: +44-113-343-4871

Academic Editor: Yulong Ding

Received: 25 January 2017; Accepted: 17 April 2017; Published: 20 April 2017

Abstract: Thermoacoustic technologies rely on a direct power conversion between acoustic and thermal energies using well known thermoacoustic effects. The presence of the acoustic field leads to oscillatory heat transfer and fluid flow processes within the components of thermoacoustic devices, notably heat exchangers. This paper outlines a two-dimensional ANSYS FLUENT CFD (computational fluid dynamics) model of flow across a pair of hot and cold heat exchangers that aims to explain the physics of phenomena observed in earlier experimental work. Firstly, the governing equations, boundary conditions and preliminary model validation are explained in detail. The numerical results show that the velocity profiles within heat exchanger plates become distorted in the presence of temperature gradients, which indicates interesting changes in the flow structure. The fluid temperature profiles from the computational model have a similar trend with the experimental results, but with differences in magnitude particularly noticeable in the hot region. Possible reasons for the differences are discussed. Accordingly, the space averaged wall heat flux is discussed for different phases and locations across both the cold and hot heat exchangers. In addition, the effects of gravity and device orientation on the flow and heat transfer are also presented. Viscous dissipation was found to be the highest when the device was set at a horizontal position; its magnitude increases with the increase of temperature differentials. These indicate that possible losses of energy may depend on the device orientation and applied temperature field.

Keywords: parallel-plate heat exchanger; standing wave; oscillatory flow; thermoacoustics

---

1. Introduction and Literature Review

Thermoacoustic systems are usually divided into engines and refrigerators depending on the direction of energy conversion between the acoustic and thermal energy. The working principle of the thermoacoustic refrigerator is illustrated in Figure 1. The acoustic driver at the end of the resonator supplies acoustic power, W_{ac}, to the working gas inside the resonator. The standing wave acoustic field induces oscillatory motion of "gas parcels" coupled with their cyclic compression and expansion. Internal structures such as heat exchangers and stacks are placed at a location where the oscillating pressure and velocity are non-zero. A fragment of a single stack plate has been magnified at the bottom of Figure 1 together with a gas parcel undergoing an acoustic oscillation. The gas parcel is compressed as it moves to the right and the parcel temperature increases to be slightly higher than the plate temperature. Consequently, heat is being transferred from the parcel to the plate. As the flow reverses, the gas parcel expands and the temperature drops slightly lower than that of the plate. As a result, heat is being absorbed from the plate and the parcel returns to its original thermodynamic state. This completes the thermodynamic cycle. In essence, the acoustic wave provides power that allows the heat to be pumped up the temperature gradient. Conversely, if a high enough temperature gradient is imposed across the stack/regenerator, an acoustic power will be self-excited and useful energy will be produced [1]. This in turn could be extracted by a linear alternator to produce electricity, or used to drive a coupled thermoacoustic cooler.

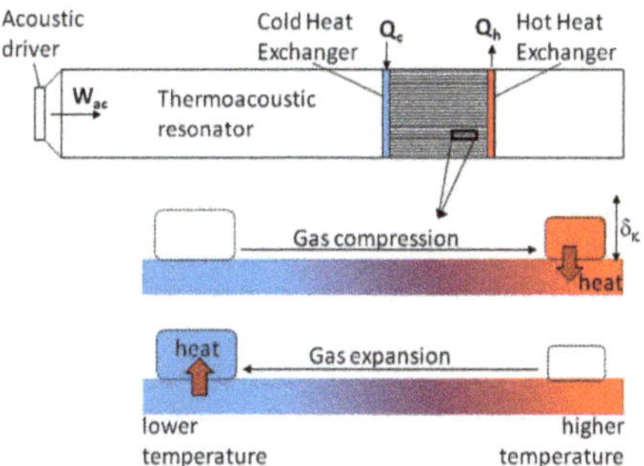

Figure 1. Schematic of a simple thermoacoustic cooler arrangement (**top**). The acoustically induced compression and expansion of fluid elements causes heat pumping effects along the stack (**bottom**).

The challenges in achieving efficient thermoacoustic devices were discussed comprehensively by Swift [1]. The difficulties with complex flows across the porous structures (stack, regenerator, and heat exchangers) that form the core of the thermoacoustic device have been raised as important research issues. Understanding phenomena such as vortex flow at the end of stacks [2], entrance effects [3], joining conditions [1], and streaming [4], is important as these could hold the key to an efficient device operation.

The structure of the vortex pattern, as observed for example by Mao et al. [5], is expected to influence the flow and possibly induce losses that may affect the efficiency of thermoacoustic systems. This has been illustrated through the observations of streamlines and viscous dissipation presented in the numerical investigation of Worlikar and Knio [6]. The viscous dissipation represents mechanical energy losses within the model. The choice of location for the stack/heat exchanger and blockage ratio (gap between plates) were reported as factors that influence mechanical energy losses.

The same group has made several improvements to their model [7–9]. These involve adiabatic stacks and heat exchangers, all working in a thermoacoustic standing wave environment. In all models, a simplified periodic configuration was used covering the area between the plates and some open area next to the plates to include the possible contribution from the vortex shedding phenomenon. The natural convection was assumed to be very small compared to the oscillatory flow magnitude, and was hence neglected. The effect of natural convection in an oscillatory flow has been previously tested in an experiment involving a heated wire located in an empty resonator [10]. The natural convection was found to dominate the flow under a small velocity regime. This result, however, is not very conclusive because the temperature difference investigated was not widely varied. Furthermore, it is speculated that the presence of additional structures such as the stack and heat exchanger may alter the influence of the natural convection effect on the flow even when the velocity is relatively small [11].

It is important to highlight that most numerical modelling work presented in this review applies a simplified model whereby a thin plate or a pair of adjacent plates with an implementation of periodic boundary conditions to replicate the array of plates is considered. In modelling, a periodic boundary condition means that the flows at the matching periodic boundary are linked so that flow conditions are shared at this boundary. As a result the flow will be duplicated for all other channels in a periodic manner [12]. While this has advantages in terms of computational cost, it must be carefully applied as it might give misleading information pertaining to the flow structure. This has been demonstrated experimentally by Guillaume and LaRue [13]. Their experimental results were later compared to a numerical model developed using commercial software, Fluent [14], using the same number of plates as in the experiment. Their works show that, for a structure with an array of plates, the flow structures at the end and in the middle of the plates are influenced by the adjacent plates. The Strouhal number (the ratio of a product of a characteristic length and the frequency of the vortex shedding to a reference velocity) is reported to have changed in magnitude when a comparison was made between the two arrangements. It is important to highlight here that the investigations reported in [13,14] were carried out in steady flow and might not be directly applicable to the oscillatory condition, but they do give an idea of possible sources of discrepancy if investigations of an array of plates are carried out using single plate approximation. The observation of asymmetry and non-periodic development of vortex structures at the end of a plate array was also reported by Mao et al. [5]. This suggests that a physical periodic structure (array of plates) does not guarantee that the flow structures are periodic. Hence modelling a single plate may not be adequate to describe the physics of flow around a structure with an array of plates.

Similarly, considering a single plate, the flow structures at two ends may not be identical in practice for reciprocating flows. Most investigations (cf., [7,9,12]) assumed that flows at the two ends are actually symmetrical and therefore the analysis was mainly focused on just one end. This seems reasonable for investigations that do not involve a temperature gradient (adiabatic stack). However, the presence of a temperature gradient on solid boundaries is necessary for the thermoacoustic effect to take place. Temperature effects must be considered to account for additional phenomena such as temperature-driven buoyancy [11,15–18] and non-linear effects [4] that could affect the flow symmetry. Experimental studies often reveal signs of nonlinearities that are insufficiently addressed by the linear model [1,11]. In most numerical studies (cf. [7,9,18]) the natural convection effect was neglected. This effect however was observed in many experimental works [1,11,17].

Current computational fluid dynamics (CFD) study aims to improve the understanding of physical processes involved in the phenomena reported in the earlier experimental work of Shi et al. [17]. However, bearing in mind more complex operating conditions (e.g., tilted-angle solar powered systems [19,20] or space applications), the scope of CFD is somewhat wider and extends to include gravity effects and effects of the device orientation on the physics of thermo-fluid processes across the heat exchanger plates. The novelty of this study stems from the fact that the model used attempts to take into account a far wider range of physical effects which are commonly neglected in numerical works related to thermoacoustics (i.e., natural convection or the use of a full array of plates to handle asymmetrical flow features). This is undertaken to explain the phenomena reported in [11,17]. In addition, the study of device orientation provides a new perspective of the importance of considering such seemingly minor details in modelling heat transfer phenomena in thermoacoustics.

2. Computational Model

A two-dimensional computational domain based on the experimental setup of [11,17] is shown by the red-dashed box in Figure 2a. It covers a rectangular area of 0.132 m $\times$ 7.4 m of the quarter-wavelength resonator. The mesh is designed to be denser within the vicinity of the heat exchanger (300 mm to the left and right from the heat exchangers) to resolve vortex structures issuing from and returning into the heat exchanger assembly. Elsewhere, a much coarser mesh is adopted as the problem consists of solving the governing equations for an empty resonator. The coarse mesh is illustrated in Figure 2b and the fine mesh in Figure 2c. This approach was validated against the theoretical predictions of linear thermoacoustic theory [1]: 300 mm away from the heat exchanger, the pressure and velocity simply follow the linear model. A structured mesh of quadrilateral type was used for this model; the long computational domain is divided into several parts so that the distribution of mesh sizes could be made with acceptable quality. The minimum orthogonal quality was recorded to be 0.74 with maximum skewness of 0.31 (i.e., close to 1 for minimum orthogonality and close to 0 for maximum skewness [15]). The minimum mesh spacing in the x- and y-directions is 0.08 mm and 0.25 mm, respectively.

The plates of the hot and cold heat exchangers (subsequently denoted as HHX for "hot heat exchanger" and CHX for "cold heat exchanger" throughout the paper) have thickness, d, of 3 mm and are arranged in a parallel configuration with the spacing, D, of 6 mm between them. This is shown in Figure 2d as a magnified view of a single channel between a pair of hot and cold plates; point 'm' is the location of the joint where the hot and cold plates meet. The HHX and CHX are located next to each other with a joint positioned at 0.17λ from the pressure antinode. The wavelength, λ, is defined as $\lambda = a/f$, where a is the speed of sound and f is the frequency of the flow (13.1 Hz). Location of the pressure antinode for this 1/4-wavelength rig is at the end wall—cf. Figure 2a. The arrangement of the heat exchanger plates (10 pairs) follows the physical setup with five pairs of heated/cooled fins and five pairs of unheated ones, treated as adiabatic plates. The unheated plates provide consistent porosities for the flow (cf., [17]).

The model extends over the whole height of the resonator to resolve the flow asymmetries observed experimentally [17]. The use of symmetry or periodic boundary conditions is inappropriate due to the presence of temperature-driven buoyancy effects. In addition, the experimental data are based on phases determined by comparing the phase of the velocity between the plates to the phase of the pressure at the pressure antinode. As the intention was to replicate as closely as possible the phenomena taking place in the physical apparatus, a full two-dimensional model was chosen.

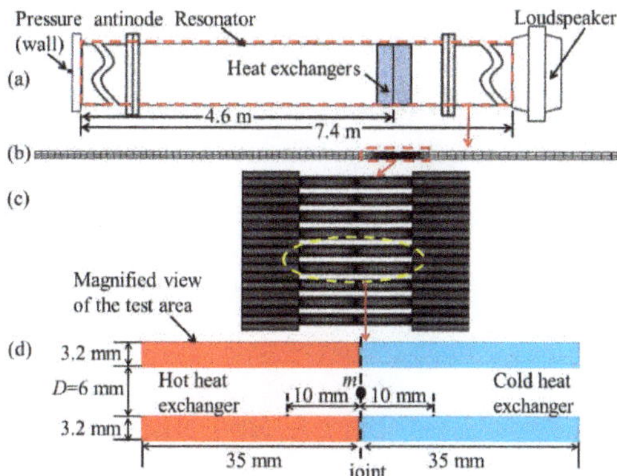

Figure 2. **(a)** A quarter-wavelength standing wave thermoacoustic rig; **(b)** computational domain of the rig covering the 7.4 m length of the resonator; **(c)** mesh generated in the area of heat exchangers and its neighbourhood; **(d)** designations/locations within the individual channel used for analysis.

The boundary condition at the right end of the resonator is defined as a moving wall to replicate the acoustic displacement induced by the loudspeaker. The displacement, δ, is simply modelled as:

$$\delta = \left[\frac{p_a}{\omega \rho_m a}\sin(k_w x_s)\right]\cos(\omega t). \quad (1)$$

Here, ω, ρ_m, a, k_w, and t are the angular velocity, mean density, speed of sound, wave number, and time, respectively. The term p_a is the oscillating pressure at the location of the pressure antinode and x_s is the distance from the joint to the pressure antinode (shown as 4.6 m in Figure 2a). The mean pressure, p_m, is 0.1 MPa. The drive ratio, p_a/p_m, of the low amplitude flow investigated throughout this paper is 0.3%. The resonator wall was treated as adiabatic. The model was solved for two thermal conditions. In the first, the walls of heat exchangers are treated as adiabatic to replicate the experiments with no temperature gradient imposed on the heat exchangers (essentially ambient laboratory conditions), mainly to make comparisons between the experimental and numerical results for vortex structures. In the second, the temperature distributions were based on experiments using a linear interpolation of experimental data reported by Shi et al. [17], as shown in Figure 3, mainly for validation of the CFD model itself. Further extended studies into the system behaviour under arbitrary conditions assumed "flat" temperature profiles (as explained later).

The flow is modelled in ANSYS FLUENT [12] using two-dimensional Navier–Stokes equations as described by:

$$\frac{\partial \rho}{\delta t} + \nabla\cdot(\rho \mathbf{v}) = 0, \tag{2}$$

$$\frac{\partial}{\partial t}(\rho \mathbf{v}) + \nabla\cdot(\rho \mathbf{v}\mathbf{v}) = -\nabla p + \nabla\cdot(\boldsymbol{\tau}) + \rho \mathbf{g}, \tag{3}$$

$$\frac{\partial}{\partial t}(\rho E) + \nabla\cdot(\mathbf{v}(\rho E + p)) = \nabla\cdot(k\nabla T + (\boldsymbol{\tau}\cdot\mathbf{v})). \tag{4}$$

The fluid used in this study is nitrogen modelled as an ideal gas with temperature-dependent properties. A power law model suggested by Swift [1] is used for the temperature-dependent viscosity while a seventh order polynomial model suggested by Abramenko et al. [21] is selected to model the temperature-dependent thermal conductivity:

$$\begin{gathered} p = \rho RT\, ; \mu = 1.82 \times 10^{-5}\left(\frac{T}{T_0}\right)^{0.69}; \\ k = -8.147 \times 10^{-4} + 1.161 \times 10^{-4}T - 1.136 \times 10^{-7}T^2 + 1.062 \times 10^{-10}T^3 - \\ 5.406 \times 10^{-14}T^4 + 1.454 \times 10^{-17}T^5 - 1.942 \times 10^{-21}T^6 + 1.011 \times 10^{-25}T^7. \end{gathered} \tag{5}$$

In Equations (2) to (5), ρ, $\mathbf{v}$, t, p, $\mathbf{g}$, E, k, T, T_0, $\boldsymbol{\tau}$, R and μ are density, velocity vector, time, pressure, gravity, energy, thermal conductivity, temperature, reference temperature (300 K), stress tensor, gas constant and viscosity, respectively [12].

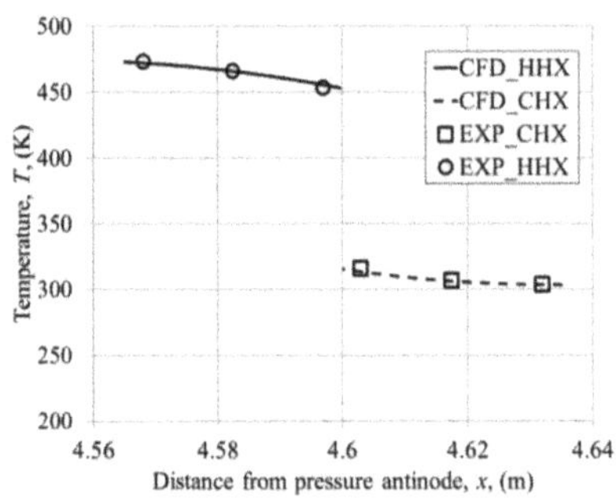

Figure 3. Cold (CHX) and hot (HHX) heat exchanger wall temperature.

Viscous heating (the second term on the right-hand-side of Equation (4)) is taken into account in this model because of the presence of heat exchangers with large surface areas. The flow is assumed to be laminar due to the small Reynolds number involved in the case investigated. The Pressure-Implicit with Splitting Operators (PISO) algorithm is selected for pressure-velocity coupling as the algorithm provides better solutions for transient cases [12,22–25].

A first order implicit scheme is used for the discretisation of time due to the presence of the moving wall [12]. The momentum and energy equations are discretised using a second order upwind scheme with the convergence set to the absolute values of 1×10^{-4} for the continuity and momentum equation, and 1×10^{-7} for energy equation. The density is calculated using the second order upwind numerical scheme. Density is related to pressure and temperature by using the equation of state as shown in Equation (5). The transient solution of the flow problem is solved in a segregated way using the pressure-based solver [12,25]. The time step size was chosen so that solution converged within 15 to 18 iterations in every time step. If the time step size was too large, the solution was found not to converge in every time step. If the size was too small, convergence occurred too fast within a certain time step (sometimes only requiring one iteration in one time step). The best time step for convergence was determined and set at 1200 steps per acoustic cycle. The area-weighted-average of pressure at the end wall, known as the pressure antinode, was monitored until a steady state oscillatory flow was obtained. This is defined as a state when pressure, velocity, and temperature do not change much from cycle to cycle. By way of example, Figure 4 shows the history of oscillating pressure and velocity monitored for location "m" as defined in Figure 2. A steady oscillatory flow condition is achieved after six flow cycles. However, as will be discussed later, results presented in this paper are obtained after 70 flow cycles, when both the flow and thermal oscillatory flow conditions have reached a steady oscillatory state. The solutions are also monitored so that they converge in every time step at every flow cycle (through the selection of the time step size).

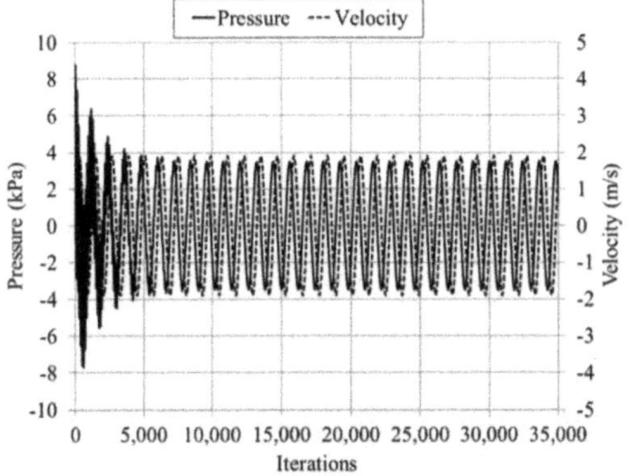

Figure 4. Oscillating pressure and velocity monitored for identification of the steady oscillatory condition.

When the steady oscillatory state condition is reached, the model is validated and used for analysis. In this study, the development of flow within one oscillatory cycle is investigated according to the time frame defined in Figure 5. The cycle is subdivided into twenty phases starting from the first phase, ϕ1. The relationship between pressure, velocity, and gas displacement, for the 20 phases of a flow cycle is illustrated. Phase ϕ1 is set for the maximum value of the oscillating pressure at the location of the pressure antinode (rigid wall of the resonator). The period of one cycle is the inverse of the operating frequency—13.1 Hz.

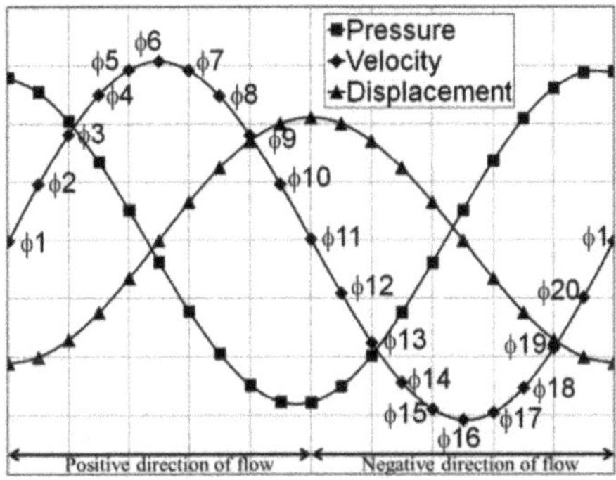

Figure 5. Pressure, velocity, and displacement over 20 phases of a flow cycle.

The model was solved using the SPECTRE High Performance Computing facility at the University of Leicester (4 login nodes for solving one model, with 28 cores of 56 threads and 256 GB memory per one login node). With this computing facility, the 2D unsteady calculation for one cycle takes about 30 min. Therefore the solution for 490 cycles (as discussed later) takes about 9.5 days.

2.1. Grid Independency

The grid independency tests are illustrated in Figure 6. In essence, the mesh was progressively refined in the part of the domain surrounding the heat exchangers (there was practically no need to do this in an empty resonator). The description of mesh density is given as cell counts in Figure 6a for five cases. Here, the velocity values at point "m" for phases 6 and 16 are taken as test variables (+ve and −ve values due to the change of direction). Similarly, the resulting axial velocity profiles (from the wall to channel centre), for selected phases in the cycle are further illustrated in Figure 6b. For clarity, velocity profiles are shown only for three cell counts. At low mesh density, the velocity profile is slightly over-predicted at all 6 phases shown. The velocity within the boundary layer appears to be the same for all cases because the mesh is always designed to be denser in that area. It is found that the model with a total of 45,910 cells is sufficient to provide a grid-independent solution.

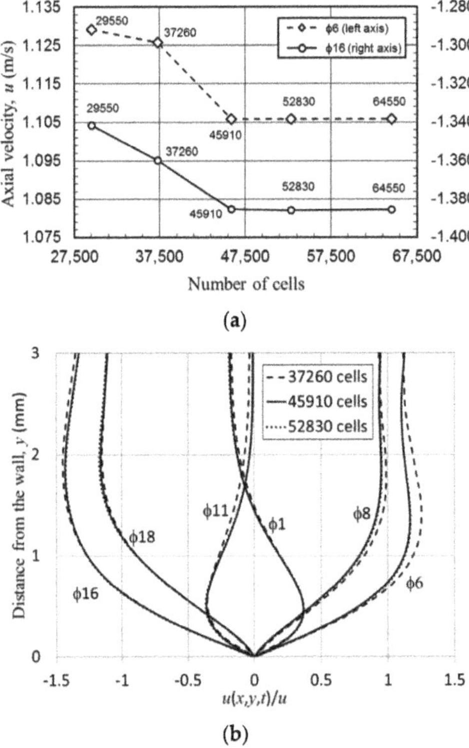

Figure 6. Grid independency test for (**a**) axial velocity amplitude at the centre of the channel for phases φ6 and φ16 and (**b**) velocity profiles near the wall; both taken 15 mm from the joint above the cold plate.

2.2. CFD Model Validation

The velocity amplitude obtained from the converged model that reaches a steady oscillatory state is compared to the experimental result to validate the model. The velocity amplitudes shown in

Figure 7a,b are for the location at the centre of the gap between the plates of the heat exchanger, 1 mm away from the joint—cf., Figure 2c—for all 20 phases in a cycle. In both graphs, the line denoted as "Theory_dT = 0" represents the theoretical/analytical solution for oscillatory velocity calculated from the general thermoacoustic theory as given by [1]:

$$u_1 = \frac{i}{\omega \rho_m}[1 - h_v(y,z)]\frac{\partial p_1}{\partial x}, \tag{6}$$

where ω, ρ_m, and $\partial p_1/\partial x$ are the angular velocity, mean density, and the pressure gradient of the oscillatory flow, respectively. The term h_v is a viscous shape factor which varies according to the geometry of the internal structure involved in the system. The shape factor for parallel plate geometry is given as [1]:

$$h_{\kappa,v} = \frac{\cosh[(1\ +\ i)y/\delta_{\kappa,v}]}{\cosh[(1\ +\ i)y_0/\delta_{\kappa,v}]}. \tag{7}$$

The shape factor defines the viscous, v, or thermal, κ, effects depending on the definition of the penetration depths used. The terms: y, y_o, and $\delta_{\kappa,v}$ are the distance from the centre of the gap to the wall of the plate, the centre of the gap, and the thermal ($\delta_\kappa = \sqrt{2\kappa/\omega}$) or viscous ($\delta_v = \sqrt{2v/\omega}$) penetration depth.

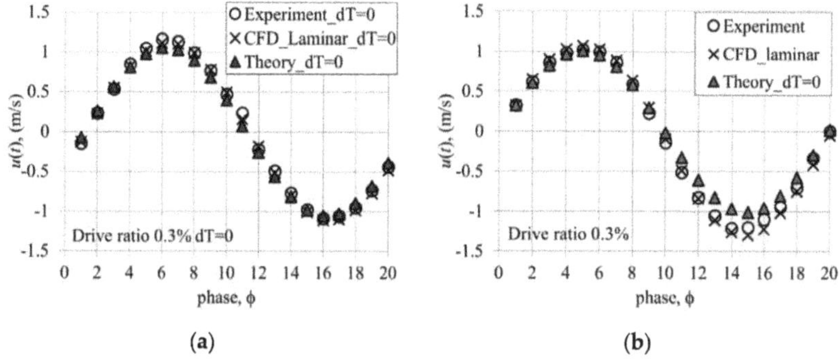

Figure 7. Centreline velocity amplitude (1 mm from the joint, above the cold plate) obtained from the model with heat exchanger walls: (**a**) adiabatic and (**b**) with the temperature profile based on the experiment.

Figure 7a, for the adiabatic case, compares the experimental results of Shi et al. [11] ("Experiment_dT = 0") with the current numerical results ("CFD_Laminar_dT = 0"). The theoretical solution is calculated for the mean density taken at 300 K. A good match between the experiment, CFD, and analytical solution can be seen. Maximum discrepancies between the three methods appear at phases ϕ6 and ϕ16 representing the highest velocity values for the two parts of the flow cycle. Comparison between the theory and experiment shows that the theoretical value is 9.3% lower than the experimental value at ϕ6, and 2.6% lower at ϕ16. However, CFD results differ from the experimental results only by 0.3% and 1.8% for ϕ6 and ϕ16, respectively.

Figure 7b shows the velocity oscillations obtained from the experiment and CFD for the case with imposed temperature difference (the theoretical adiabatic solution is only plotted for reference). Experimental and CFD results show that, with the imposed wall temperature condition, the flow in the second half of the cycle exhibits a larger amplitude of velocity compared to the first half of the cycle. This is counterintuitive since one would expect that a hotter fluid would travel with higher velocity in the first half of the cycle, while cold fluid would flow with a lower velocity in the second half of the cycle. However it is possible that there is a mean flow along the channels (streaming effect) due to the convective currents as discussed later in Section 3.2.2.

Figure 8 shows the validation of temperature profiles for a location 15 mm from the joint—cf., Figure 2c—above both the cold and hot plates. The maximum relative discrepancies of 6% were found between the numerical and experimental values when using the absolute temperature scale (17 K on the HHX and 18 K on CHX). However these discrepancies need to be seen in the context of high experimental uncertainty of the temperature field when using acetone based planar laser induced fluorescence (PLIF) method (16 K reported in [17]) as well as a large temperature span between the heated and cooled plates—cf. Figure 3. Clearly, the trend of the temperature profiles agrees very well qualitatively with the experimental values. Indeed, the original PLIF experiments were more concerned with temperature gradients than absolute temperature values in order to calculate surface heat fluxes.

Of course, even if the temperature measurement had zero error, there would still inevitably be discrepancies in CFD arising from the inability to model all aspects of the physical process. For example, in reality the walls of the rig are not adiabatic, and in addition they do accumulate heat and tend to induce thermal inertia processes with the scale of hours. The role of the three-dimensional flow effects within the system is also unclear—unfortunately, the experiment could only provide "planar" results close to the resonator centreline.

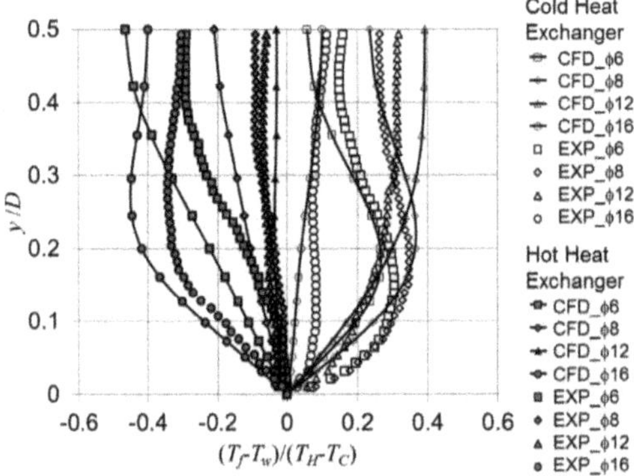

Figure 8. Temperature profiles for HHX and CHX—comparison between experimental (EXP) and computational (CFD) results at the location 15 mm from the joint (above the HHX and CHX plates).

3. Results and Discussions

Results presented in this paper look at four aspects of CFD modelling: the choice of temperature field initialisation, the effects of wall temperature on the flow field, device orientation relative to gravity force and viscous dissipation.

3.1. *Investigation of the Effect of the Initial Fluid Temperature on Flow and Heat Transfer*

The current CFD study deals with flow fields with relatively high imposed temperature gradients which results in difficulties to achieve numerically a temperature field similar to that obtained from the experiment. A possible way of bringing these close to each other, within a reasonable computational time, was through setting an appropriate initial fluid temperature in the model. Three approaches were investigated and these are discussed below. Their outcomes are illustrated for phase $\phi 1$ in Figure 9; the experimental data shown in Figure 9a are compared against CFD data in Figure 9b–d.

First, the model was initialised with a uniform fluid temperature, $T_i = 300$ K. The model was iterated until it reached 70 flow cycles which is equivalent to 5 s of real time. Figure 9b shows that the resulting temperature field is still dominated by the low temperature of the initial 300 K. On the other hand, the experimental data in Figure 9a show that the average fluid temperature is considerably higher due to the fact that the experimental data was collected after the flow has conveniently settled in the steady oscillatory state. At that stage, the heat has been accumulated in the system giving rise to the mean temperature of the flow. For the computational model with initial temperature of 300 K, a significant amount of computational time and effort would be required if a similar heat accumulation as in the experiment were to be achieved.

Second, the model was initialised with a uniform fluid temperature close to the mean temperature from the experiments, $T_i = 360$ K and was similarly iterated for 70 cycles. Although the resulting temperature field in Figure 9c is closer to the experimental results, the effect of natural convection in the open area next to the CHX is not visible in the computational result. In addition, the temperature of the fluid within the channel of the CHX reduces from the initial high temperature to a lower temperature due to the cooling effect at that location. The reason for the mismatch is not clear, especially as the wall of the heat exchanger is already modelled following the temperature measured by the thermocouples as reported in the experimental work [17]. However, it is possible that additional features need to be introduced into the model such as heat losses (the modelled resonator is adiabatic, in reality there are considerable heat losses), heat accumulation, or heat leakages. The known issues of experimental temperature uncertainties need to be borne in mind too.

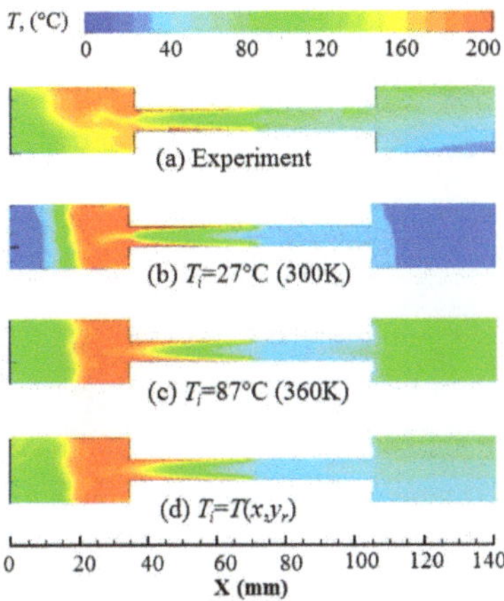

Figure 9. Temperature contours at phase ϕ1 from the experiment (**a**) and three numerical models initialised with different temperature fields (**b–d**).

The third approach tried to improve the predictions of natural convection observed experimentally at the plate ends by initialising the model with an approximate temperature field estimated from the experimental results ($T_i \approx T(x, y_r)$). The resulting temperature contour after 70 cycles is shown in Figure 9d. The effect of the initialisation temperature seems to dominate the temperature field especially at the open area next to the plates. In the area between the plates of the heat exchanger, the wall temperature seems to control the temperature within the area.

Figure 10 shows the effect of initial temperatures on the velocity amplitudes at a location 1 mm away from the joint above the cold plate. Evidently, a higher mean temperature of 360 K results in a slightly higher amplitude of the velocity, especially during the second half of the cycle. This may suggest that the experimental temperature field is affected by the accumulation of heat within the investigated area since raising the initial temperature from 300 to 360 K brings the solution closer to the experimental results. The same observation can be made for all locations along the heat exchanger plate as illustrated in Figure 10b. The velocity amplitudes at ϕ6 and ϕ16 shown in Figure 10b represent the highest velocity amplitude during the first and second parts of the flow cycle. The velocity amplitude for the model initialised at 360 K is always higher than the other two models for all the locations along the heat exchanger plate.

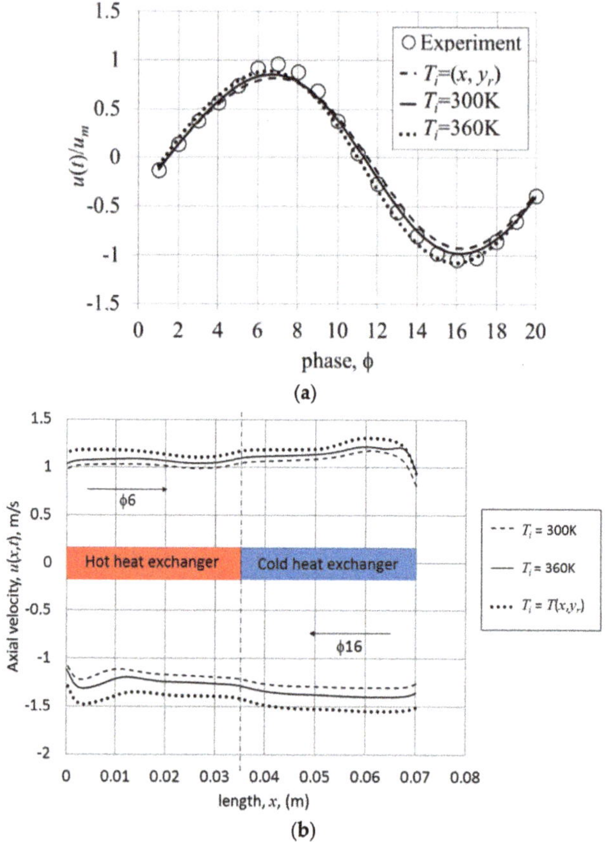

Figure 10. The effect of initial temperature on (**a**) the oscillating velocity at location 1 mm from the joint above the cold plate and (**b**) the velocity amplitude along the heat exchanger's plate taken at the centre of the channel for ϕ6 and ϕ 16.

Finally, the effect of initial temperature on the heat flux is shown in Figure 11. Here, the heat flux is spatially averaged over the length of the heat exchanger (the same way as in [8]), and is presented as a function of time (phase in the cycle):

$$q_{h,c} = \frac{1}{l}\int_0^l -k\frac{dT}{dy}\bigg|_{wall} dx. \tag{8}$$

Subscripts h and c refer to HHX and CHX, respectively. The results were taken at cycle 70 after the model reached a steady oscillatory condition.

Overall, the space-averaged wall heat fluxes from CFD are in-phase with the experimental values at both heat exchangers. However the magnitude differs at the HHX, especially when the flow starts slowing down at ϕ6 and changes direction. The numerical space-time-averaged wall heat flux is calculated to be 30% lower than the experimental values. However, the difference is within the range reported in other numerical investigations [26]. The wall heat flux predicted by the model initialised at 360 K seems closest to the experiment. Also, the predictions for CHX seem to follow the experiment better. The above discussion shows that the choice of initial temperature field is an important factor in the model, and it may lead to difficulties with convergence and accuracy.

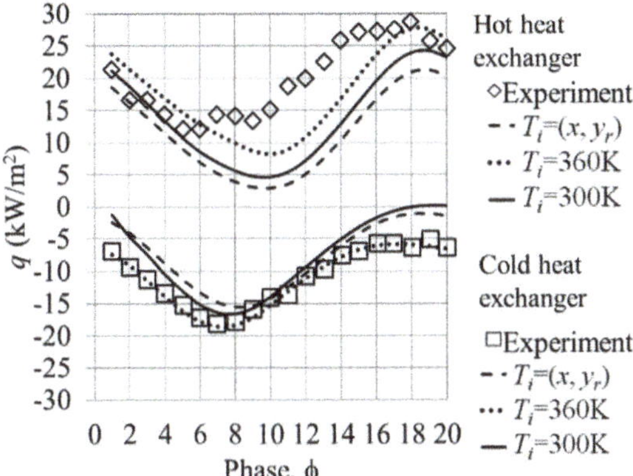

Figure 11. The effect of initial temperature on the wall heat flux of the heat exchangers.

Figure 12 shows the temperature profiles plotted along the vertical direction y_r, between the bottom and top wall of the resonator, at locations in the open areas: 40 mm from the cold end and 38 mm from the hot end of the heat exchangers, that is where most of natural convection effects are identified experimentally. Clearly, initialising the temperature at 360 K (dotted lines) gives the wrong values of temperature especially at the lower area of the domain compared to the experiments. Initialising the temperature field as $T(x, y_r)$ (dashed lines), seems to give a closer match in this area.

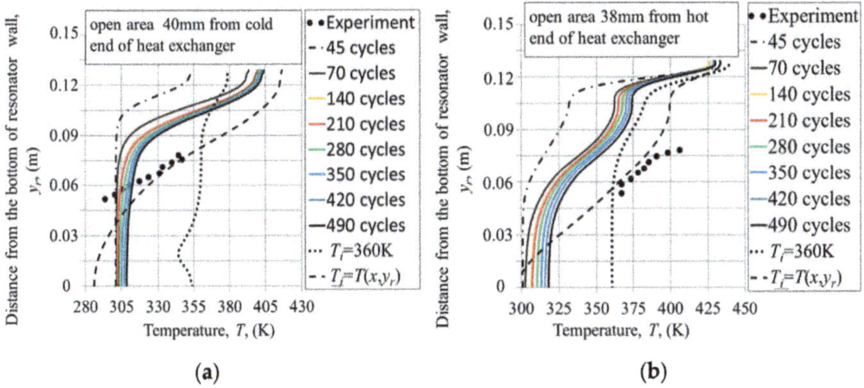

Figure 12. Temperature profiles at the open area next to the (**a**) cold and (**b**) hot ends of the heat exchanger assembly—comparison between the experimental and numerical model with different approaches of temperature initialisation.

Finally, it is interesting to look at the development of the temperature field as a function of the number of cycles taken in the solution. Here the model initialised at 300 K is taken as an example; the temperature profiles are plotted in Figure 12 for increasing the total number of cycles as coloured lines between the dash-dotted black line (45 cycles) and solid black line (490 cycles). The temperature profiles seem to tend towards the experimental values. However, the changes become smaller and smaller (especially after about 70 cycles) and it is unlikely that the profiles would ever reach the experimental data. Following this observation, all the models used in the analysis reported in this paper were run for at least 70 cycles so that the steady oscillatory thermal condition is reached from a practical point of view, in addition to the steady oscillatory flow condition reported in Figure 4.

Figure 13 shows the temperature field within and around the heat exchanger assembly for the initial temperature of 300 K after 490 cycles. This illustrates the convective currents at both the hot and cold ends to supplement Figure 12. The heat accumulation phenomena shown in Figure 13 were also seen in [8]. The results illustrate a potential source of streaming in thermoacoustic systems [1] caused by buoyancy driven flows, which may interfere with the main thermoacoustic processes.

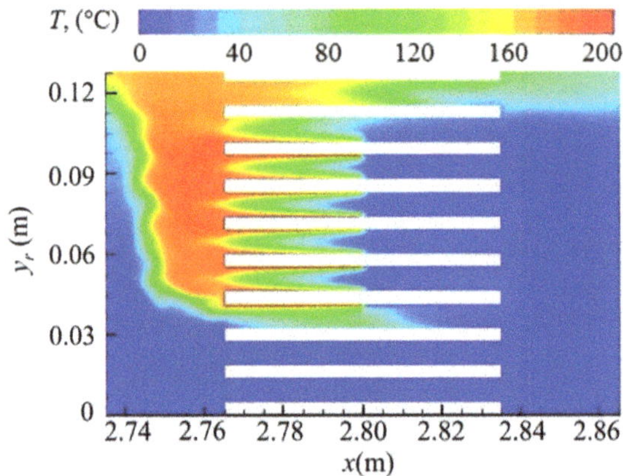

Figure 13. Illustration of the convective currents occurring at the top area of the heat exchanger assembly.

Figure 14 shows the temperature profiles plotted from the heat exchanger wall to $y = D/2$ (centre of the channel between the plates) at a location 10 mm from the joint above both the hot and cold plates. These are plotted for selected phases in the cycle (ϕ6, ϕ8, ϕ12, and ϕ16) for the experiment and three temperature initialisation cases, after 70 cycles. The top four plots are for the cold plate; the bottom four plots are for the hot plate. An initial temperature of 360 K leads to profiles with magnitudes larger than the experiment and it was already shown that it models the natural convection in the open areas poorly. The use of temperature distribution $T(x, y_r)$ provides solutions almost similar to a model initialised at 300 K. Both offer temperature profiles closer to the experiment. However, from the practical perspective of experimental accuracies, it would be hard to obtain the actual experimental temperature distributions for the sake of CFD, and therefore all subsequent models developed in this study are initialised at 300 K and iterated for 70 cycles.

Figure 15a shows the experimental temperature contours for all twenty phases of the cycle inside the channel formed by the HHX and CHX walls, while Figure 15b shows the CFD solutions for the same "viewing area" as used in the experiment. Qualitatively, there seems to be a fairly good agreement between the two in terms of flow physics. The differences in magnitude, particularly at the right side of the viewing area, are mainly due to the difficulty of achieving numerically the experimentally recorded mean temperature of the flow. However the trends of the temperature distributions are the same for all phases in the cycle. Hot fluid starts flowing into the channel from the left during the first part of the cycle (ϕ1–ϕ10). As the flow reverses (ϕ11–ϕ20), the cold fluid starts flowing from the right. The temperature within the channel is bounded by the experimental temperature profile set at the cold plate (cf. Figure 3). Therefore achieving higher mean temperature within the cold channel, as observed in the experiment, cannot be obtained under the current numerical time integration and modelling. Of course the discrepancies between CFD and the experiments may come from a variety of sources including the already discussed experimental errors as well as incorrect assumptions behind the numerical model compared to reality, e.g., the already discussed adiabatic boundary conditions, heat accumulation (vs. leakage) problems, or unknown (and experimentally unverifiable) three-dimensional effects.

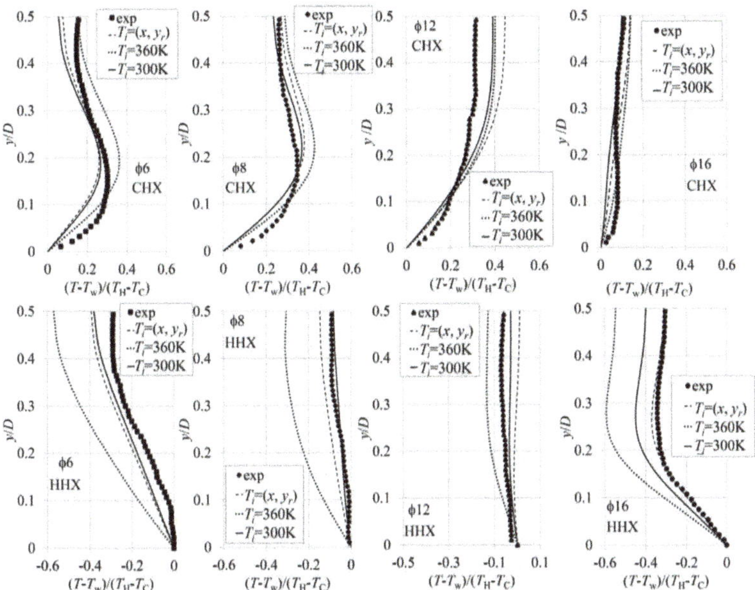

Figure 14. The effect of initial temperature on the temperature profiles between the heat exchanger plates at a location 10 mm from the joint above the cold (CHX), and hot (HHX) plates.

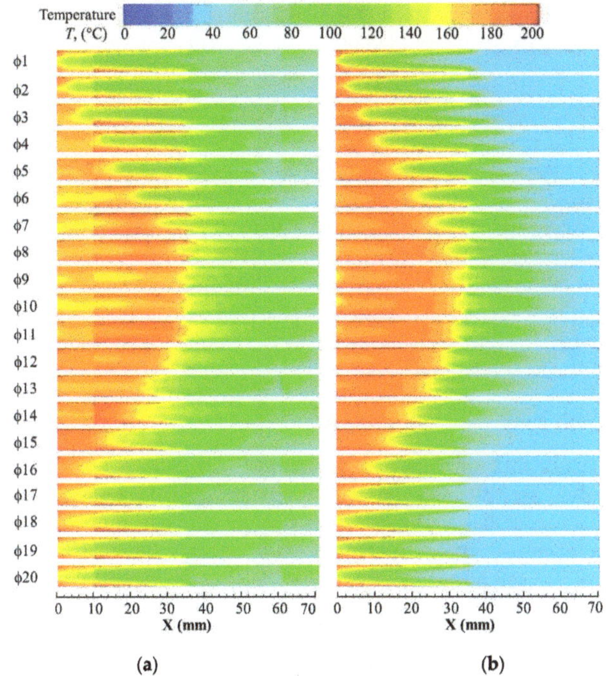

Figure 15. Temperature contours in the area bounded by the heat exchanger walls—comparison between (**a**) experiment of Shi et al. [17] and (**b**) simulation.

3.2. The Effect of the Heat Exchanger Wall Temperature on the Flow Field

The presence of the temperature field introduces new elements to the flow physics including natural convection, stronger forced convection due to gas thermal expansion, or thermo-viscous interplay in the boundary layers. A few different sets of thermal boundary conditions were applied to heat exchangers, as summarised in Table 1. Here, the temperature profile was assumed to be "flat" as only one case had experimental data available. However, this particular case (T_H = 200 °C, T_C = 30 °C) was solved in two ways: first, with the wall temperature from Figure 3, for code validation, and second, with the "flat" temperature profiles to compare between different numerical cases. The drive ratio (defined as the ratio of pressure at the antinode to the mean pressure) of the flow is maintained at 0.3%.

Table 1. Wall temperature condition of the heat exchanger.

Drive Ratio (%)	Working Medium	Frequency (Hz)	Mean Pressure (bar)	Heat Exchanger Wall Temperature (°C)	
				Hot, T_H	**Cold, T_C**
				Adiabatic	Adiabatic
				200	30
0.3	Nitrogen	13.1	1	100	30
				300	30
				300	100

3.2.1. Study of Flow Using the Adiabatic Model

The investigation of flow across the parallel plate structure starts with a model where the walls of the heat exchangers are treated as adiabatic. The resulting velocity profiles plotted at a location of 1 mm from the joint above the cold plate are shown in Figure 16 for comparison to the experimental results of Shi et al. [11]. The agreement with the experimental data is very good. Figure 17 shows the contours of vorticity, ω_z, plotted for selected phases of the oscillatory flow, obtained as:

$$\omega_z = \frac{\partial v}{\partial x} - \frac{\partial u}{\partial y}, \tag{9}$$

where u and v are the velocity components in the x and y directions of the flow. The contours are plotted within the same "viewing area" as in the experiment [11]—cf. also Figure 2d. For phases ϕ1 to ϕ10, the fluid is flowing into the channel from the left; for phases ϕ11 to ϕ20 from the right. At the beginning of the cycle, a pair of small vortices appear at the entry to the channel followed by a pair of stronger counter-rotating vortices, which appear to "leap-frog" the original pair around phase ϕ4. However, both vortex structures dissipate by about phase ϕ8. A similar feature is present at the other end of the channel when the flow changes direction. The laminar model captures the features of the oscillatory flow at 0.3% drive ratio very well.

3.2.2. The Effect of Wall Temperature on the Flow and Heat Transfer

The effect of temperature on the flow and heat transfer is investigated by setting the heat exchanger wall temperatures to the values shown in Table 1. It should be noted that all comparisons to experiments were made using the model that applies the experimental temperature profiles of Figure 3 on the hot and cold plates (of course, except cases marked as "adiabatic" or "dT = 0"). However, for the sake of comparisons between numerical cases, "flat" temperature profiles were set up, as already mentioned in the introduction to Section 3.2. Figure 18 shows the comparison between the experimental and numerical velocity profiles for the flow at the location of 10 mm from the joint above the cold plate for comparable temperature gradients set up along the plates. The slight differences in the magnitude between the experimental and CFD data can be accounted for by the typical errors of particle image velocimetry (PIV), although it is worth mentioning that thermophoresis which affects seeding particles in the presence of temperature gradients [27] can also have an additional impact on the experimental data.

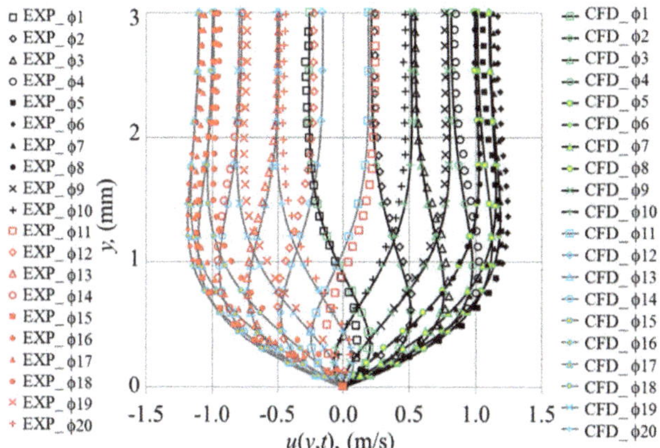

Figure 16. Velocity profiles from computational fluid dynamics (CFD) model (black and grey lines) and the experiment (red and black symbol) for all 20 phases of a flow cycle. The heat exchanger walls are adiabatic.

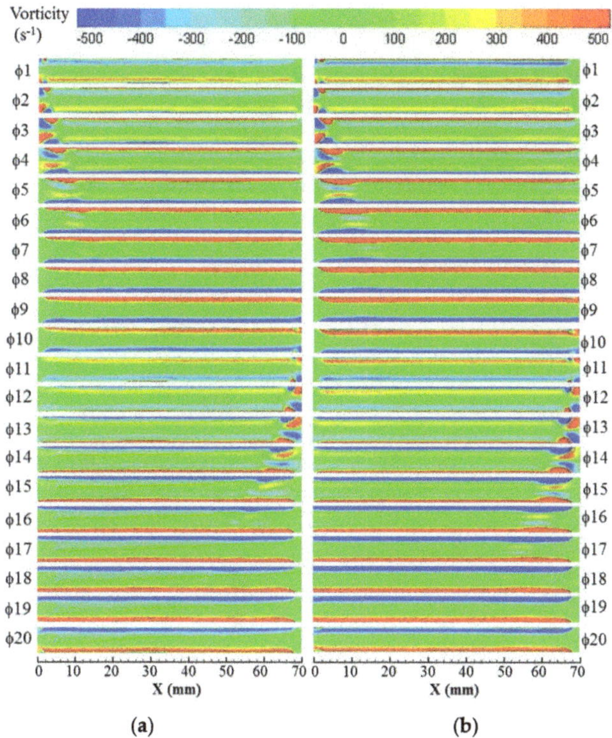

Figure 17. Vorticity contours within the channel with plates treated as adiabatic walls—comparison between the results from (a) the experiment of Shi et al. [11] and (b) CFD.

As shown in Figure 9a, buoyancy effects are observed in the experimental results. The heat supplied by the HHX causes reduced density of the fluid, which results in a buoyancy driven flow in the open area next to the HHX superimposed on the forced convection due to the flow being ejected from the channel by acoustic excitation as well as thermal expansion of the fluid travelling in the hot part of the channel. These temperature effects are likely contributors to the flow asymmetries observed in Figure 7b. These effects are also seen through a comparison of velocity profiles between the models with and without the temperature gradient imposed on the heat exchanger walls (cf., Figure 16 vs. Figure 18).

The above asymmetry can be observed in two ways: Firstly, the whole structure of the velocity profile plotted in Figure 18 seems to be "leaning" to the left in comparison to Figure 16. This is clearly illustrated in Figure 19, where the presence of temperature makes the velocity profile of ϕ9 and its counterpart in reversed flow, ϕ19, "lean" to the left.

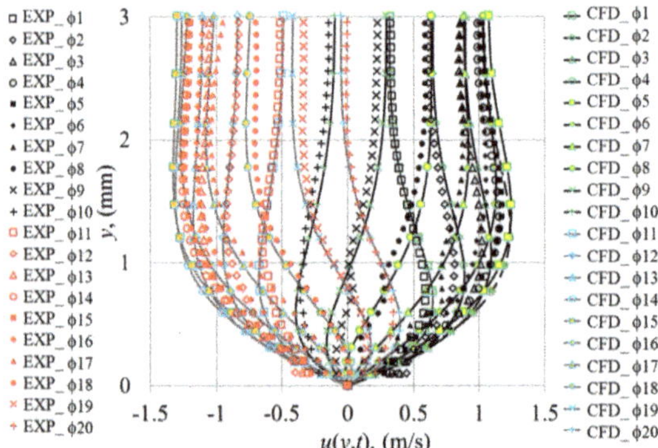

Figure 18. Velocity profiles from CFD (black and grey lines) and the experiment (red and black symbol) for 20 phases of a flow cycle with the influence from heat exchanger wall temperature.

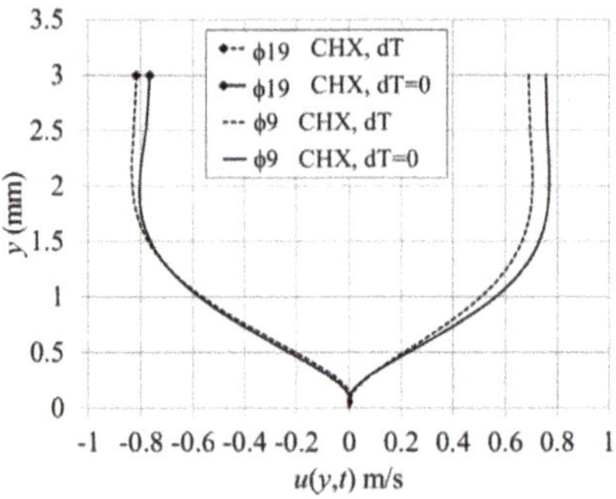

Figure 19. The effect of the imposed temperature gradient (dT) on the velocity profile taken at 15 mm from the joint above the cold plate (CHX).

Secondly, the maximum magnitude of velocity fluctuation extends further to the centreline when temperature effects are included in the model as seen in Figure 20a. In the absence of temperature effects (adiabatic wall condition shown in Figure 20 as dT = 0), the velocity profile is similar at both CHX and HHX. When the temperature effect is considered, the velocity profile at the HHX seems shifted up towards the centreline (y = 3 mm) and the magnitude of the velocity is larger compared to an equidistant location at the CHX. On the other hand, focusing on one location, Figure 20b shows that the velocity profiles at the selected point of the CHX in analogous phases of the two flow directions (i.e., ϕ9 vs. ϕ19 here), are also affected by the presence of temperature effects. For ease of comparison, the axial velocity of ϕ19 in Figure 20b is presented as a negative (−ve) value. The presence of temperature effects results in a lower centreline velocity at ϕ9 but bigger magnitudes in the reversed flow direction at ϕ19. The results indicate the changes in the viscous boundary layer due to the presence of temperature.

The asymmetry of the flow structure can be further illustrated through the vorticity contours at the open area next to the heat exchanger for three different wall temperature conditions (Figures 21–23). The vorticity is plotted for two phases corresponding to the opposite flow direction (ϕ1 and ϕ11). For each phase, the contour is shown with two views of the two ends of the heat exchanger assembly.

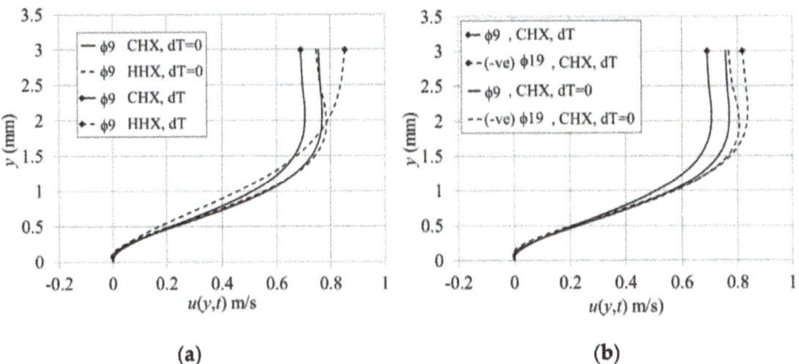

Figure 20. Velocity profile; (**a**) for ϕ9 at CHX and HHX, both at a location 15 mm away from the joint; (**b**) at CHX only at the different flow direction.

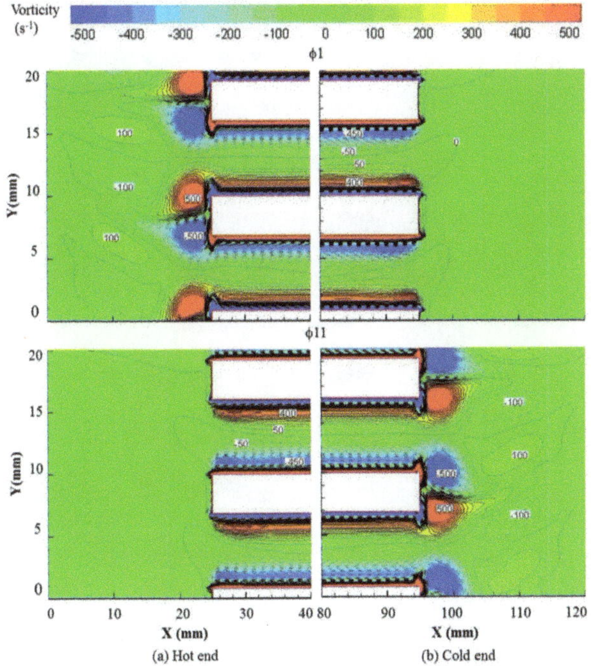

Figure 21. Vorticity contour for the case of the heat exchanger set as the adiabatic wall at the (**a**) hot and (**b**) cold plates. (1 bar in nitrogen gas, 0.3% drive ratio, 13.1 Hz).

Figure 21 illustrates the case of adiabatic conditions and it is clear that the flow around two ends of the heat exchanger assembly is symmetrical, without any noticeable buoyancy driven convective patterns. Figure 22 shows the structure of the vortices when there is a temperature difference imposed: 200 °C at the hot plate, and 30 °C at the cold plate. The symmetry of the flow is broken: as the viscosity of the gas increases with temperature, the thicknesses of the viscous penetration depth increases, as seen, for example, in different separations of the vorticity contours on the right and left ends of the channel, which is also congruent with the velocity profiles in Figure 19a. Consequently, the vortex structures rolling up on the two sides of the plate assembly differ in size and strength. However, in addition, the temperature-driven flow at the open area at the left introduces some differences in the vortex patterns between the top and bottom side of the plate—note the hot plume observed in Figures 9 and 13 which adds a "cross-flow" (vertically upward) component relative to the hot fluid ejected from the channels. Interestingly Figure 22a, for phase $\phi 11$, shows the artefacts of a vortex structure on the top of the hot plate, but not on the bottom. These flow features are strengthened in Figure 23, for temperature conditions of 300 °C at the hot, and 30 °C at the cold plates.

There are two more effects that could be speculated on: First, that the large vortex structures created on the left side of hot plates may provide additional resistance to the flow when it reverses and starts to move from left to right by "bottlenecking" the channel. The same effect may be at play on the right side of the CHX when the flow reverses and starts moving to the left, but such flow would only need to handle much smaller vortices. Second, the temperature driven flow in the open area to the left of the hot plates may lead to setting up a variety of convective "cells" (loops), which could be contained on one side of the plate assembly, but equally induce a local streaming current between the plates (some of these being also adiabatic plates above and below the heated/cooled channels). Both of these effects could be further contributors to breaking the flow symmetry in the acoustic cycle.

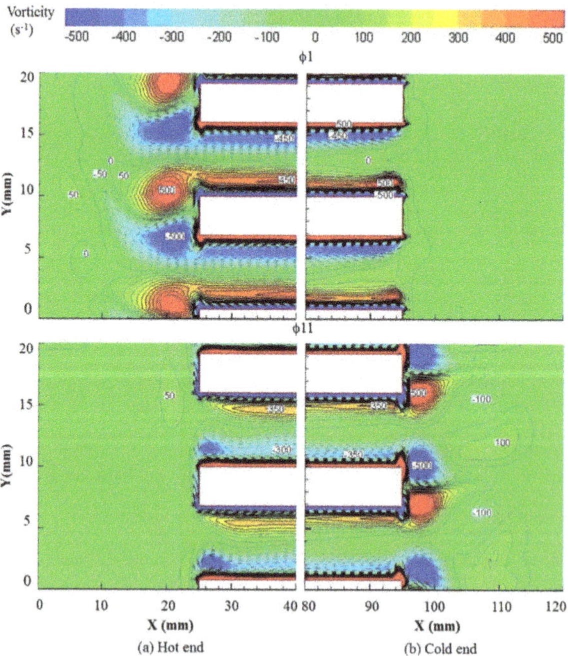

Figure 22. Vorticity contour for the case of the heat exchanger set at 200 °C at the (**a**) hot and 30 °C at the (**b**) cold plates. (1 bar in nitrogen gas, 0.3% drive ratio, 13.1 Hz, temperature profiles assumed "flat").

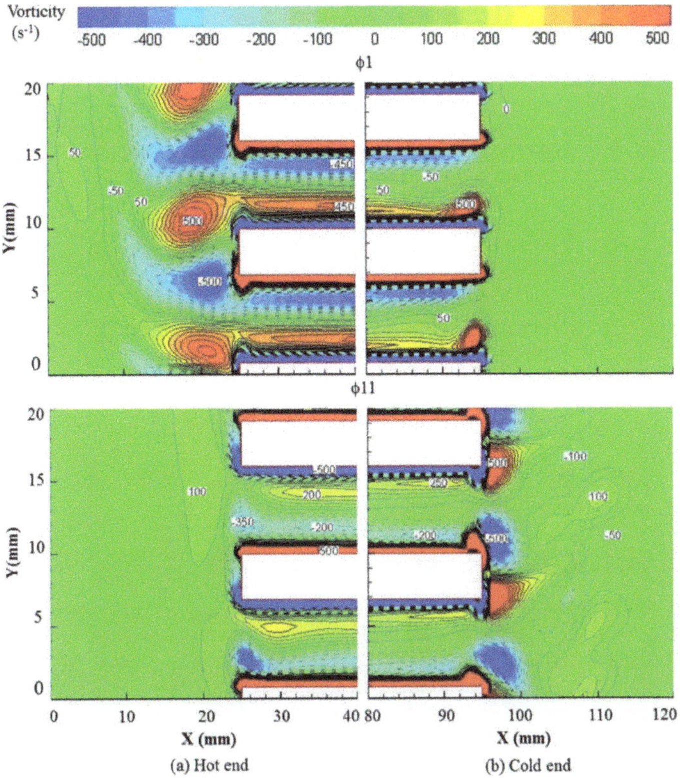

Figure 23. Vorticity contour for the case with the heat exchanger set at 300 °C at the (**a**) hot and 30 °C at the (**b**) cold plates. (1 bar in nitrogen gas, 0.3% drive ratio, 13.1 Hz, temperature profiles assumed "flat").

Figure 24 summarizes all studies with the varied temperature difference between the hot and cold plates (cf., Table 1) in terms of the centreline velocity amplitude along the length of the heat exchanger channel. Two selected phases, ϕ9 and ϕ19, representing the first half (flow to the right) and the second half of the cycle (flow to the left) are chosen, respectively; the legend with thermal conditions is shown at the top. The variation of the plots with varying thermal conditions is indicative of a very complex physics with a number of effects at play. Clearly, the variation of the channel wall temperature and corresponding gas expansion (or contraction) causes gas particles to travel to a different distance back and forth during its reciprocating movement, causing asymmetrical velocity amplitudes on the centreline. However, the exact shape of the velocity profile depends on thermal (and thus viscous) conditions across the oscillatory boundary layers causing velocity "overshoots" and "deficits". These may correspond to the counterintuitive reduction in the centreline velocity for the cold channel in phase ϕ9 when the case (T_H = 100 °C, T_C = 30 °C) is replaced by cases (T_H = 200 °C, T_C = 30 °C) and (T_H = 300 °C, T_H = 30 °C) and an increase for case (T_H = 300 °C, T_C = 100 °C). Similar counterintuitive phenomena can be seen in other places of Figure 24 for both selected phases. The already mentioned convection "cells" appearing across the overall domain and the varying size and structure of vortices rolling up on the plate ends (and subsequently sucked back into the channels) may also need to be considered as contributors to the appearance of plots in Figure 24.

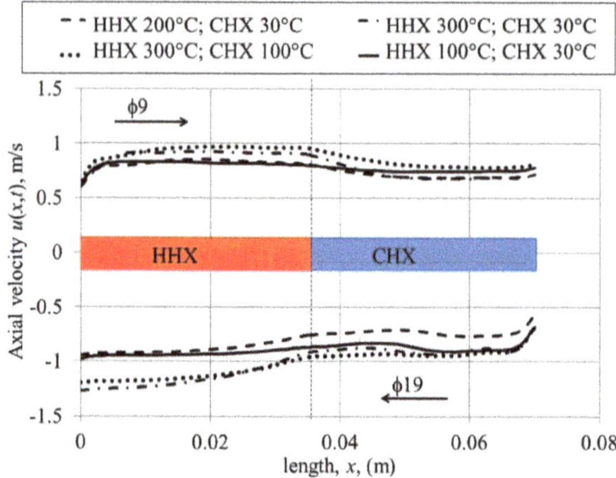

Figure 24. Axial velocity at the middle of the channel along the hot (HHX) and cold (CHX) heat exchangers; temperature profiles assumed "flat".

3.3. *The Effect of Gravity and Device Orientation on Flow and Heat Transfer*

The orientations of the resonator, relative to the gravity field, investigated in this study are schematically illustrated in Figure 25 (the three pairs of plates are for illustration only—all ten pairs are included in the model as before). The direction of the gravity field is modelled accordingly following the orientation. 0° corresponds to the horizontal layout in the figure, 90° indicates the hot plates above the cold plates, and 270° indicates the hot plates below the cold plates. The case with gravity switched off (g = 0) is also considered for reference. Experimental temperature profiles shown in Figure 3 are applied at the cold and hot plates in the models.

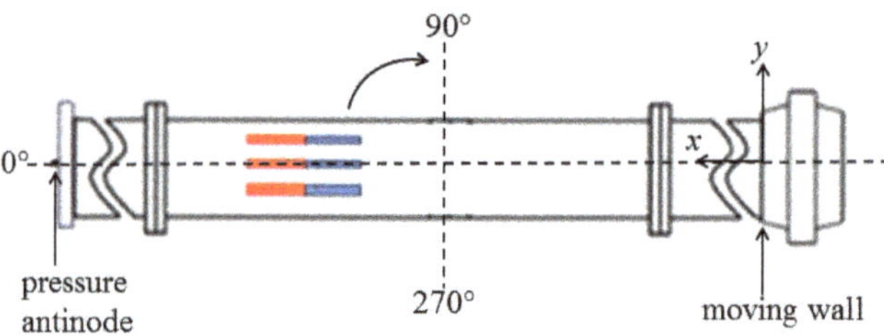

Figure 25. Illustration of the thermoacoustic device orientation. Red and blue plates symbolise hot and cold heat exchangers, respectively.

Figure 26 shows the effect of gravity and device orientation (tilt angle 0°, 90°, and 270°) on the velocity profile plotted 15 mm from the joint above the cold and hot plates for selected phases ϕ9 and ϕ19. In order to differentiate between the data for the cold and hot channels, the lines for the velocity profiles at the hot channel are assigned a "diamond" symbol at the top end of the lines. The effect of gravity can be seen by comparing cases assigned as "g, 0°" (presented by solid-lines) and "g = 0" (presented by dashed-lines). For phase ϕ9 (Figure 26b), the comparison shows that gravity has little effect on the flow structure in the horizontal orientation. However, the effect of gravity is seen to be larger at ϕ19. Comparison between the magnitude of the velocity between ϕ9 and ϕ19 for both "g, 0°" and "g = 0" shows that the temperature-driven buoyancy flow seems to cause a stronger asymmetry between the first and second parts of the flow cycle. The temperature-driven flow due to gravity seems to significantly assist the flow at ϕ19 and slightly resist the flow at ϕ9.

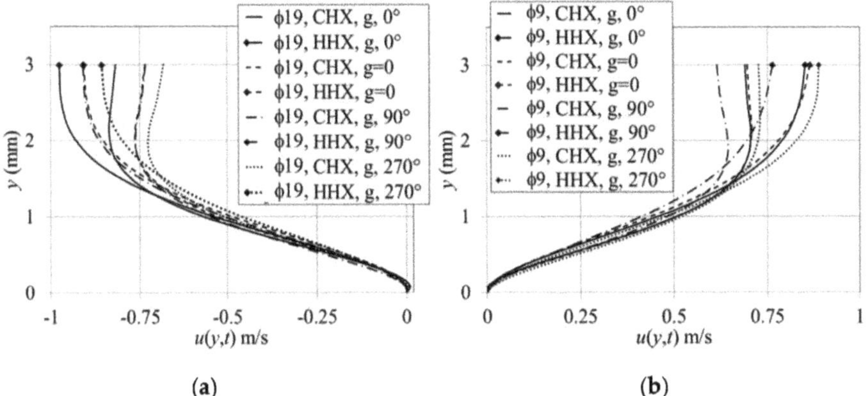

Figure 26. The effect of gravity and device orientation on the velocity profiles at a location 15 mm from the joint above the cold (CHX) and hot (CHX) plates for phases (**a**) ϕ19 and (**b**) ϕ9.

For an orientation other than horizontal, the gravity field affects the flow depending on the direction of flow and the relative locations of the HHX and CHX. At an orientation of 90° (presented by the dashed-dotted lines), the cold plate is located below the hot plate resulting in a reduced velocity magnitude during ϕ9 of the flow cycle, because the buoyancy effects are resisting the flow. As the flow reverses (shown as ϕ19), the buoyancy effect is helping the flow. As a result, the velocity magnitude becomes bigger than that at the first half of the cycle.

Conversely, when the device is at an orientation of 270° (presented by the dotted-lines), the temperature-driven buoyancy effect assists the flow at ϕ9 (Figure 26b), resulting in a higher velocity amplitude and hence a greater fluid displacement between the plates. At ϕ19 (Figure 26a), the flow reverses and the temperature-driven buoyancy effect is resisting the flow. Consequently, the velocity amplitude within the channel is lower than that in the first part of the cycle. For all device orientations, the velocity amplitude within the hot channel is always bigger than that at the cold channel. This is a logical consequence of the temperature-dependent density as previously discussed.

The effect of gravity and device orientation on the wall heat transfer is shown in Figure 27. The heat flux is averaged over the length of the heat exchanger and is calculated using Equation (8). The effect of gravity is less pronounced in the CHX compared to the HHX. This is in accordance with the effect of buoyancy driven flow, which is more pronounced at the HHX compared to the CHX. These effects can be looked at from the point of view of temperature profiles shown in Figure 28. It can be seen that the orientation changes the temperature field depending on the location of the HHX, the direction of flow, and the resulting temperature-driven buoyancy effect. When the flow is moving in the positive direction (ϕ9), the effect of device orientation on the temperature profiles is more pronounced at the CHX. When the flow reverses (ϕ19), the effect appears to be more obvious at the location of the HHX. Natural convection effect at ϕ19 for HHX is stronger than that at ϕ9 for CHX because the 'hot plume' appears next to the HHX (as seen in Figure 13). This is congruent with the weak gravity effects on the heat flux at the CHX compared to the HHX as seen in Figure 27. Of course, at an orientation of 270° (dashed line), when the HHX is located below the CHX, the temperature between both the HHX and CHX increases due to the temperature-driven buoyancy effect. At an orientation of 90°, presented by a dashed-dotted line, the temperature between the plates and hence the wall heat flux is lower because heat is driven away from the plates of the heat exchanger by the buoyancy effect.

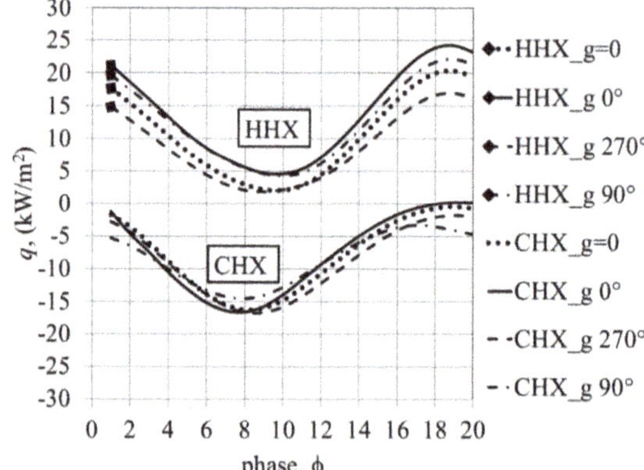

Figure 27. The effect of gravity and device orientation on the wall heat flux of the hot (HHX) and cold (CHX) plates.

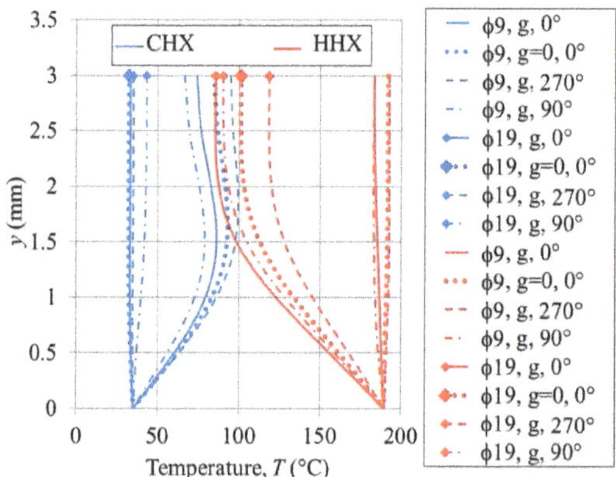

Figure 28. Temperature profiles for two selected phases, when there is no gravity (g = 0) and with gravity, g, at different tilt angles; 0°, 90°, 270°, at a location 15 mm from the joint above the cold (CHX) and hot (HHX) plates.

3.4. Viscous Dissipation

Typically, it is assumed that viscous dissipation effects play a small role for low speed flows. However, in the flow situations presented in this paper, the contribution of viscous dissipation due to the existence of internal structures within the investigated domain may be significant and hence it is considered here. Dimensional analysis presented in Appendix A reveals the dimensionless parameters that need to be considered. In particular, it is found that comparison between viscous dissipation in areas of empty resonator and heat exchanger plates should take into consideration porosity, σ, and Eckert number, Ec—cf., Equation (A7).

In a viscous flow, the energy of the fluid motion (kinetic energy) is transformed into the internal energy of the fluid through the existence of viscous dissipation. Viscous dissipation for a two-dimensional model is defined as [28]:

$$\Phi = 2\left\{\left(\frac{\partial u}{\partial x}\right)^2 + \left(\frac{\partial v}{\partial y}\right)^2\right\} + \left(\frac{\partial v}{\partial x} + \frac{\partial u}{\partial y}\right)^2 - \frac{2}{3}\left(\frac{\partial u}{\partial x} + \frac{\partial v}{\partial y}\right)^2. \quad (10)$$

The computational domain is divided into four regions defined in Figure 29. The normalised viscous dissipation is averaged over the cycle and space, A, corresponding to each region and calculated as:

$$\langle \Phi^* \rangle = \frac{1}{2\pi A} \int_0^A \int_0^{2\pi} (\mu^* \Phi^*) d\phi dA. \tag{11}$$

The term in the bracket on the right-hand-side of Equation (11) is calculated as follows:

$$\mu^* \Phi^* = \frac{1}{\mu_c} \left(\frac{d_c}{u_c \sigma} \right)^2 \mu \left[2 \left\{ \left(\frac{\partial u}{\partial x} \right)^2 + \left(\frac{\partial v}{\partial y} \right)^2 \right\} + \left(\frac{\partial v}{\partial x} + \frac{\partial u}{\partial y} \right)^2 - \frac{2}{3} \left(\frac{\partial u}{\partial x} + \frac{\partial v}{\partial y} \right)^2 \right]. \tag{12}$$

Following the dimensional analysis, the weighted-average value is further multiplied by the dimensionless parameters identified as porosity, σ, and Eckert number, Ec:

$$\langle \Phi \rangle = \sigma^2 Ec \langle \Phi^* \rangle. \tag{13}$$

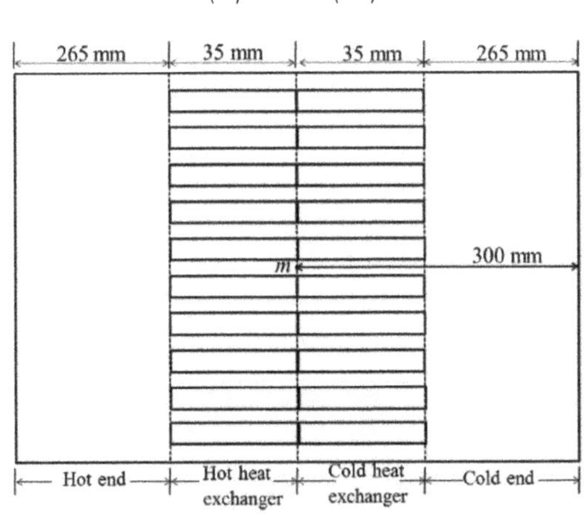

Figure 29. Illustration of the region defined for the analysis of viscous dissipation.

The reference velocity is taken at 300 mm from the joint of the heat exchangers (labelled as m in Figures 2 and 29) towards the right end of the figure within the region called the "cold end". The reference location is selected to avoid the influence of temperature on the open area next to the HHX, where a hot plume is observed. A value of porosity, σ, is appropriately used when calculating dissipation for the open area and the area of the heat exchanger following the earlier discussion. For the open area, the porosity is set to one while an appropriate value of porosity (0.65 for the geometry investigated) is used for the heat exchangers.

Figure 30 shows the effect of the heat exchanger wall temperature on the dimensionless viscous dissipation. The viscous dissipation at the open areas at both ends of the heat exchangers are one order of magnitude lower than the dissipation in the area containing the plates. This is consistent with the fact that the working medium interacts with more wall surfaces within the area of the heat exchanger where viscous resistance is significant, hence causing more energy dissipation. When the HHX becomes hotter, the dissipation becomes bigger. This is a consequence of gas viscosity that becomes bigger as the temperature increases. Figure 30b has an expanded y-axis to show dissipation levels at the ends of the heat exchanger assembly. It shows that the dissipation of energy at the open area next to the hot end is bigger than that next to the cold end and that it increases with an increase of temperature of the HHX. This is likely the combined effect of the larger gas viscosity at higher temperature and the more pronounced gravity-driven flow as seen by the development of the hot plume at that end.

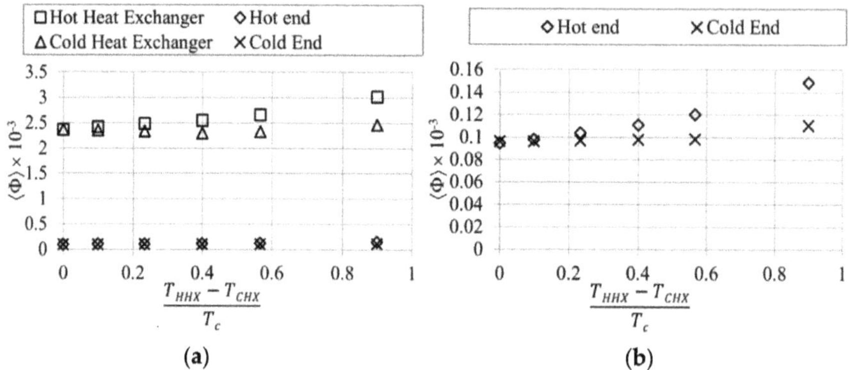

Figure 30. (**a**) The effect of the heat exchanger wall temperature on viscous dissipation and (**b**) the enlarged view for dissipations at the open area next to the heat exchangers.

Figure 31 shows that viscous dissipation is lower in all areas when the device is set at a vertical orientation. Gravity determines the flow direction of the hot plume. For vertical arrangements (90° and 270°), the temperature-driven flow is in line with the direction of the flow. The temperature-driven flow can either assist or obstruct the flow depending on the device orientation and the direction of flow within a flow cycle. This may affect the magnitude of the velocity as the fluid flows back and forth within the heat exchanger, but the gradients of velocity near the wall are approximately the same, as seen in Figure 26. However, the low dissipation of the device arranged at vertical orientation indicates that the combined effect of temperature-driven flow and the mean flow may help reduce the viscous dissipation that may relate to the re-circulation of the fluid. The horizontal arrangement (0°) results in a temperature-driven flow perpendicular to the direction of the flow. Hence, re-circulation tends to be bigger in comparison to the vertical arrangement.

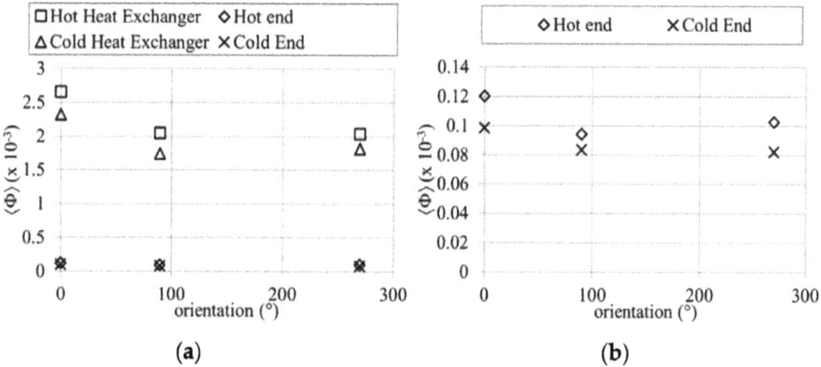

Figure 31. (**a**) The effect of the device's orientation on viscous dissipation and (**b**) the enlarged view for viscous dissipation at the open area next to the heat exchangers.

4. Conclusions

The two-dimensional ANSYS FLUENT CFD model has been set up to investigate the oscillatory flow across the parallel-plate heat exchanger at the low drive ratio of 0.3%. The flow, as represented by the velocity profiles, is well predicted by the model. For example, the velocity amplitudes at the centre of the heat exchanger channel have a maximum discrepancy of 1.8% compared to the experimental data. On the other hand, the numerically calculated temperatures have the maximum discrepancies of 6% compared to the experimental values. The differences noted in the temperature profiles and the resulting heat fluxes are likely to be the result of the combination of factors related to experimental uncertainties and the inability of the model to capture all the complexities of the real experimental setup. Overall, the current model under-predicts the space-time-averaged wall heat transfer by up to 30% in comparison to the experimental values.

The investigation of the effect of initial temperature on the flow and heat transfer helps explain the selection of an appropriate initial temperature that best suits the model. The current approaches of prescribing the initial temperature fields showed that the initialisation procedure influences both the magnitude of heat flux and velocity profiles within the structures. The investigation suggests that the current model is best initialised at 300 K. The model initialised with a temperature field defined more closely to the experimental conditions resulted in a change of the flow character, incongruent with the real flow behaviour. It can be speculated that the reason for this is the type of boundary conditions imposed on the model, which may not fully represent the reality of the experiment with its long time-history (hours), far beyond the capabilities of the current CFD model.

The imposed temperature boundary conditions (on the heat exchanger plates) is shown to cause asymmetry to the flow seemingly related to the nonlinearity caused by the combined effects of the temperature-dependent density and viscosity and the temperature-driven flow (buoyancy). The change in fluid density with respect to the temperature results in lower velocity amplitudes during the first part of the cycle and relatively higher magnitudes of velocity for the second part of the cycle. Here, an "asymmetrical boundary" creates asymmetries in the temperature and velocity fields. In addition, the temperature-driven buoyancy flow creates further "distortions" to the already asymmetrical features of the velocity and temperature fields. The results indicate that there are "net flows" through channels (or in other words "streaming"). These effects are well illustrated through changing the orientation of the device relative to the gravity field.

Temperature-dependent gravity effects influence the magnitudes of velocity profiles and temperature profiles between the plates depending on the direction of the flow and the locations of the CHX/HHX. Consequently, the heat fluxes also change. The presence of the imposed temperature field is shown to influence the velocity profile and gas displacement across the plates, hence breaking the symmetry of the flow. The vortex structures at the end of the plates, and the shear layer within the area bounded by the plates, change with temperature. Device orientation influences the flow and heat transfer due to temperature-driven buoyancy effects.

The viscous dissipation is also affected. The vertical arrangement is shown to provide a lower viscous dissipation but care should also be taken when dealing with this kind of arrangement because the temperature-driven flow changes the temperature and velocity field within the channel.

Acknowledgments: The first author would like to acknowledge the sponsorship of the Ministry of Education, Malaysia and Universiti Teknikal Malaysia Melaka (FRGS/1/2015/TK03/FKM/03/F00274). The second author would like to acknowledge the sponsorship of the Royal Society Industry Fellowship (2012–2015), grant number IF110094 and EPSRC Advanced Research Fellowship, grant GR/T04502.

Author Contributions: Fatimah A.Z. Mohd Saat performed the numerical investigations, analysed the data, and drafted the paper; Artur J. Jaworski planned and supervised the research work, outlined the paper, and subsequently worked with first author on consecutive "iterations" of the paper until the final manuscript was produced.

Conflicts of Interest: The authors declare no conflict of interest. The funding sponsors had no role in the design of the study; in the collection, analyses, or interpretation of data; in the writing of the manuscript, and in the decision to publish the results.

Appendix A. Dimensional Analysis

The dimensional analysis required in Section 3.4 is based on the analysis of the transport equations that govern the flow. The analysis is carried out using a two-dimensional Navier–Stokes equation for compressible flow. In this study, the normalisation parameters are introduced as:

$$\begin{aligned} u^* &= \frac{u}{\sigma u_c};\ v^* = \frac{v}{\sigma u_c};\ x^* = \frac{x}{d_c};\ y^* = \frac{y}{d_c};\ t^* = t\omega;\ p^* = \frac{p}{\rho_c \sigma^2 u_c^2}; \\ \rho^* &= \frac{\rho}{\rho_c};\ g^* = \frac{g}{g_c};\ \beta^* = \frac{\beta}{\beta_c};\ \mu^* = \frac{\mu}{\mu_c};\ T^* = \frac{T - T_c}{T_H - T_c};\ c_p^* = \frac{c_p}{c_{p,c}};\ k^* = \frac{k}{k_c}. \end{aligned} \tag{A1}$$

The superscript * describes the dimensionless variable. The subscript c refers to a reference value, preferably selected in the open area a distance away from the stack/regenerator where the properties and behaviour of the fluid are not affected much by the presence of the object. The velocity is normalised by the velocity at the reference point multiplied by the porosity, σ. The dimensions x and y are both normalised by the characteristic dimension, d_c—here the length of the heat exchanger plate (35 mm) was used. The temperatures T_H and T_C selected represent the hot and cold plate/fluid temperatures, respectively.

The resulting dimensionless equations for the two-dimensional Navier–Stokes equations are:
Continuity equation

$$Re_\omega \frac{\partial \rho^*}{\partial t^*} + \sigma Re_d \frac{\partial(\rho^* u^*)}{\partial x^*} + \sigma Re_d \frac{\partial(\rho^* v^*)}{\partial y^*} = 0, \tag{A2}$$

x-momentum equation

$$\begin{aligned} &Re_\omega \frac{\partial}{\partial t^*}(\rho^* u^*) + \sigma Re_d \frac{\partial}{\partial x^*}(\rho^* u^{*2}) + \sigma Re_d \frac{\partial}{\partial y^*}(\rho^* u^* v^*) + \sigma Re_d \frac{\partial p^*}{\partial x^*} = \\ &\frac{Gr}{\sigma Re_d}\rho^* g^* \beta^* T^* + \frac{\partial}{\partial x^*}\left[2\mu^* \frac{\partial u^*}{\partial x^*} - \frac{2}{3}\mu^* \left(\frac{\partial u^*}{\partial x^*} + \frac{\partial v^*}{\partial y^*}\right)\right] + \frac{\partial}{\partial y^*}\left[\mu^* \left(\frac{\partial v^*}{\partial x^*} + \frac{\partial u^*}{\partial y^*}\right)\right], \end{aligned} \tag{A3}$$

y-momentum equation

$$\begin{aligned} &Re_\omega \frac{\partial}{\partial t^*}(\rho^* v^*) + \sigma Re_d \frac{\partial}{\partial x^*}(\rho^* u^* v^*) + \sigma Re_d \frac{\partial}{\partial y^*}(\rho^* v^{*2}) + \sigma Re_d \frac{\partial p^*}{\partial y^*} = \\ &\frac{Gr}{\sigma Re_d}\rho^* g^* \beta^* T^* + \frac{\partial}{\partial y^*}\left[2\mu^* \frac{\partial v^*}{\partial y^*} - \frac{2}{3}\mu^* \left(\frac{\partial u^*}{\partial x^*} + \frac{\partial v^*}{\partial y^*}\right)\right] + \frac{\partial}{\partial x^*}\left[\mu^* \left(\frac{\partial v^*}{\partial x^*} + \frac{\partial u^*}{\partial y^*}\right)\right], \end{aligned} \tag{A4}$$

Energy equation

$$\begin{aligned} &Re_\omega \theta \frac{\partial}{\partial t^*}\left(\rho^* c_p^* T^*\right) + \sigma Re_d \theta \left[\frac{\partial}{\partial x^*}\left(\rho^* c_p^* u^* T^*\right) + \frac{\partial}{\partial y^*}\left(\rho^* c_p^* v^* T^*\right)\right] = \frac{\theta}{Pr}\left[\frac{\partial}{\partial x^*}\left(k^* \frac{\partial T^*}{\partial x^*}\right) + \frac{\partial}{\partial y^*}\left(k^* \frac{\partial T^*}{\partial y^*}\right)\right] + \\ &\sigma^2 Ec\left[Re_\omega \frac{\partial p^*}{\partial t} + \sigma Re_d \left(\frac{\partial}{\partial x^*}(p^* u^*) + \frac{\partial}{\partial y^*}(p^* v^*)\right)\right] + \sigma^2 Ec(\mu^* \Phi^*). \end{aligned} \tag{A5}$$

The term Φ^* is defined as:

$$\Phi^* = 2\left\{\left(\frac{\partial u^*}{\partial x^*}\right)^2 + \left(\frac{\partial v^*}{\partial y^*}\right)^2\right\} + \left(\frac{\partial v^*}{\partial x^*} + \frac{\partial u^*}{\partial y^*}\right)^2 - \frac{2}{3}\left(\frac{\partial u^*}{\partial x^*} + \frac{\partial v^*}{\partial y^*}\right)^2, \tag{A6}$$

The dimensionless parameters discovered through this procedure are:

$$\begin{aligned} Re_\omega &= \frac{\omega d_c^2}{v_c};\ Re_d = \frac{u_c d_c^2}{v_c};\ Gr = \frac{g_c \beta_c (T_H - T_c) d_c^3}{v_c^2};\ \theta = \frac{T_H - T_C}{T_c}; \\ Ec &= \frac{u_c^2}{c_{p,c} T_c};\ Pr = \frac{\mu_c c_{p,c}}{k_c} \end{aligned} \tag{A7}$$

The kinetic Reynolds number, Re_ω, describes the influence of frequency on the flow. The hydraulic Reynolds number, Re_d, describes the flow based on the amplitude of the velocity, u_c. The combined effect of frequency and flow amplitude is related to a dimensionless number known as the Keulegen-Carpenter number, $KC = Re_d/Re_\omega$. The temperature-driven characteristic is represented by the Grashof number, Gr. The dimensionless temperature, θ, represents the effect of temperature on the flow. Depending on the definition of the reference temperature used, this dimensionless number may represent the effect of heat accumulation within the investigated area. The effect can be seen in the transient and convection part of the energy equation. This indicates that heat accumulation changes over time and is very much dependent on the amplitude of the flow. The Eckert number, Ec, expresses the relationship between the flow kinetic energy and enthalpy, and is used to characterise dissipation. The Prandtl number, Pr, describes the property of fluids, which is useful when comparing cases with different fluid media.

References

1. Swift, G.W. *Thermoacoustics: A Unifying Perspective for Some Engines and Refrigerators*; Acoustical Society of America Publications: Sewickley, PA, USA, 2002.
2. Mao, X.; Jaworski, A.J. Oscillatory flow at the end of parallel-plate stacks: Phenomenological and similarity analysis. *Fluid Dyn. Res.* **2010**, *42*, 055504. [CrossRef]
3. Jaworski, A.J.; Mao, X.; Mao, X.; Yu, Z. Entrance effects in the channels of the parallel plate stack in oscillatory flow conditions. *Exp. Therm. Fluid Sci.* **2009**, *33*, 495–502. [CrossRef]
4. Matveev, K.; Backhaus, S.; Swift, G. On Some Nonlinear Effects of Heat Transport in Thermal Buffer Tubes. In Proceedings of the 17th International Symposium on Nonlinear Acoustics, Pennsylvania, PA, USA, 18–22 July 2006.
5. Mao, X.; Yu, Z.; Jaworski, A.J.; Marx, D. PIV studies of coherent structures generated at the end of a stack of parallel plates in a standing wave acoustic field. *Exp. Fluids* **2008**, *45*, 833–846. [CrossRef]
6. Worlikar, A.S.; Knio, O.M. Numerical simulation of a thermoacoustic refrigerator I. Unsteady adiabatic flow around the stack. *J. Comput. Phys.* **1996**, *127*, 424–451. [CrossRef]
7. Worlikar, A.S.; Knio, O.M.; Klein, R. Numerical simulation of thermoacoustic refrigerator II. Stratified flow around the stack. *J. Comput. Phys.* **1998**, *144*, 299–324. [CrossRef]
8. Besnoin, E.; Knio, O.M. Numerical Study of thermoacoustic heat exchangers in the thin plate limit. *Numer. Heat Transfer. Part A* **2001**, *40*, 445–471.
9. Besnoin, E.; Knio, O.M. Numerical study of thermoacoustic heat exchangers. *Acta Acust. United Acust.* **2004**, *90*, 432–444.
10. Mozurkewich, G. Heat transfer from a cylinder in an acoustic standing wave. *J. Acoust. Soc. Am.* **1995**, *98*, 2209–2216. [CrossRef]
11. Shi, L.; Yu, Z.; Jaworski, A.J. Application of laser-based instrumentation for measurement of time-resolved temperature and velocity fields in the thermoacoustic system. *Int. J. Therm. Sci.* **2010**, *49*, 1688–1701. [CrossRef]
12. *ANSYS Fluent*; 13.0; User Manual; ANSYS Inc.: Canonsburg, PA, USA, 2010.
13. Guillaume, D.W.; LaRue, J.C. Comparison of the vortex shedding behaviour of a single plate and plate array. *Exp. Fluids* **2001**, *30*, 22–26. [CrossRef]
14. Guillaume, D.W.; LaRue, J.C. Comparison of the numerical and experimental flow field downstream of a plate array. *J. Fluids Eng.* **2002**, *124*, 184–186. [CrossRef]
15. Parang, M.; Salah-Eddine, A. Thermoacoustic Convection Heat-Transfer Phenomenon. *AIAA* **1984**, *22*, 1020–1022.
16. Lin, Y.; Farouk, B.; Oran, E.S. Interactions of thermally induced acoustic waves with buoyancy induced flows in rectangular enclosures. *Int. J. Heat Mass Transf.* **2008**, *51*, 1665–1674. [CrossRef]
17. Shi, L.; Mao, X.; Jaworski, A.J. Application of planar laser-induced fluorescence measurement techniques to study the heat transfer characteristics of parallel-plate heat exchangers in thermoacoustic devices. *Meas. Sci. Technol.* **2010**, *21*, 115405. [CrossRef]
18. Kuzuu, K.; Hasegawa, S. Numerical investigation of heated gas flow in a thermoacoustic device. *Appl. Therm. Eng.* **2017**, *110*, 1283–1293. [CrossRef]
19. Shen, C.; He, Y.; Li, Y.; Ke, H.; Zhang, D.; Liu, Y. Performance of solar powered thermoacoustic engine at different tilted angles. *Appl. Therm. Eng.* **2009**, *29*, 2745–2756. [CrossRef]
20. Yassen, N. Impact of temperature gradient on thermoacoustic refrigerator. *Energy Procedia* **2015**, *74*, 1182–1191. [CrossRef]

21. Abramenko, T.N.; Aleinikova, V.I.; Golovicher, L.E.; Kuz'mina, N.E. Generalization of experimental data on thermal conductivity of nitrogen, oxygen, and air at atmospheric pressure. *J. Eng. Phys. Thermophys.* **1992**, *63*, 892–897. [CrossRef]
22. Tasnim, S.H.; Fraser, R.A. Computation of the flow and thermal fields in a thermoacoustic refrigerator. *Int. Commun. Heat Mass Transf.* **2010**, *37*, 748–755. [CrossRef]
23. Cha, J.S.; Ghiaasiaan, S.M. Oscillatory flow in microporous media applied in pulse-tube and Stirling-cycle cryocooler regenerators. *Exp. Therm. Fluid Sci.* **2008**, *32*, 1264–1278. [CrossRef]
24. Frederix, E.M.A.; Stanic, M.; Kuczaj, A.K.; Nordlund, M.; Geurts, B.J. Extension of the compressible PISO algorithm to single-species aerosol formation and transport. *Int. J. Multiph. Flow* **2015**, *74*, 184–194. [CrossRef]
25. Versteeg, H.K.; Malalasekera, W. *An Introduction to Computational Fluid Dynamics the Finite Volume Method*, 2nd ed.; Prentice Hall: Essex, UK, 2007; pp. 179–197, 263–265.
26. Piccolo, A. Numerical computation for parallel plate thermoacoustic heat exchangers in standing wave oscillatory flow. *Int. J. Heat Mass Transf.* **2011**, *54*, 4518–4530. [CrossRef]
27. Talbot, L.; Cheng, R.K.; Schefer, R.W.; Willis, D.R. Thermophoresis of particles in a heated boundary layer. *J. Fluid Mech.* **1980**, *101*, 737–758. [CrossRef]
28. Schlichting, H.; Gersten, K. *Boundary Layer Theory, 8th revised and enlarged edition*; Springer: New Delhi, India, 2001; pp. 76, 416–507.

© 2017 by the authors. Licensee MDPI, Basel, Switzerland. This article is an open access article distributed under the terms and conditions of the Creative Commons Attribution (CC BY) license (http://creativecommons.org/licenses/by/4.0/).

Article

Numerical Predictions of Early Stage Turbulence in Oscillatory Flow across Parallel-Plate Heat Exchangers of a Thermoacoustic System

Fatimah A. Z. Mohd Saat [1] **and Artur J. Jaworski** [2,*]

[1] Centre for Advanced Research on Energy, Faculty of Mechanical Engineering, Universiti Teknikal Malaysia Melaka, Hang Tuah Jaya, Durian Tunggal, Melaka 76100, Malaysia; fatimah@utem.edu.my

[2] Faculty of Engineering, University of Leeds, Woodhouse Lane, Leeds LS2 9JT, UK

* Correspondence: a.j.jaworski@leeds.ac.uk; Tel.: +44-113-343-4871

Academic Editor: Yulong Ding

Received: 19 April 2017; Accepted: 31 May 2017; Published: 30 June 2017

Abstract: This work focuses on the predictions of turbulent transition in oscillatory flow subjected to temperature gradients, which often occurs within heat exchangers of thermoacoustic devices. A two-dimensional computational fluid dynamics (CFD) model was developed in ANSYS FLUENT and validated using the earlier experimental data. Four drive ratios (defined as maximum pressure amplitude to mean pressure) were investigated: 0.30%, 0.45%, 0.65% and 0.83%. It has been found that the introduction of the turbulence model at a drive ratio as low as 0.45% improves the predictions of flow structure compared to experiments, which indicates that turbulent transition may occur at much smaller flow amplitudes than previously thought. In the current investigation, the critical Reynolds number based on the thickness of Stokes' layer falls in the range between 70 and 100. The models tested included four variants of the RANS (Reynolds-Averaged Navier–Stokes) equations: k-ε, k-ω, shear-stress-transport (SST)-k-ω and transition-SST, the laminar model being used as a reference. Discussions are based on velocity profiles, vorticity plots, viscous dissipation and the resulting heat transfer and their comparison with experimental results. The SST-k-ω turbulence model and, in some cases, transition-SST provide the best fit of the velocity profile between numerical and experimental data (the value of the introduced metric measuring the deviation of the CFD velocity profiles from experiment is up to 43% lower than for the laminar model) and also give the best match in terms of calculated heat flux. The viscous dissipation also increases with an increase of the drive ratio. The results suggest that turbulence should be considered when designing thermoacoustic devices even in low-amplitude regimes in order to improve the performance predictions of thermoacoustic systems.

Keywords: parallel-plate heat exchanger; oscillatory flow; standing wave; thermoacoustic system; turbulence; transition

1. Introduction

Oscillatory wall-bounded flow emerges in a variety of engineering applications, one of the good examples being thermoacoustic devices. These are known as a technology that offers greener alternatives for energy conversion applications in areas such as waste heat recovery, solar power or environmentally-friendly cooling technologies. One of the keys to a better design of thermoacoustic systems lies in the understanding of the flow and heat transfer within the internal structures in oscillatory conditions. Experimental studies often reveal signs of nonlinearities that are insufficiently addressed by the linear models typically used for design predictions [1,2]. These may be a result of phenomena such as acoustic streaming, vortex shedding or turbulence in the flow [3–5].

Previous studies of oscillatory flows identified many interesting features of the transition process [6–9]. The resulting turbulent flow is characterised by the sudden, explosive appearance of turbulence towards the end of the acceleration phase of the cycle. Turbulent flow is sustained throughout the deceleration phase, while during the early stages of the acceleration phase, the production of turbulence essentially stops, and velocity profiles agree with laminar theory [8]. Nevertheless, during this period, the disturbances retain a small, but finite energy level.

Theoretical [1,10] and numerical [11–15] works related to thermoacoustic processes typically assume a laminar flow (limited to low drive ratios). The critical Reynolds numbers for the transition to turbulence in an oscillatory flow are reviewed by Ohmi et al. [6]. Their experimental results agree with the critical value of 400 of the Reynolds number defined by Merkli and Thomann [7] based on the Stokes layer thickness. The flow regions are categorised into laminar, transitional and turbulent. However, oscillatory flows of a Reynolds number higher than 400 are shown experimentally to experience a stage of relaminarization where the velocity profiles match the laminar prediction during the initial stage of the acceleration phase and change to turbulent-like profiles at the later phase, before turning back to laminar [8,9].

An additional complication in the transition/relaminarization processes lies in the appearance of temperature-dependent fluid properties, compressibility effects or additional forces, such as gravity [8]. Therefore, necessary precautions need to be taken in the modelling of such flow. As for the size of the computational domain, Fieldman and Wagner [13] suggested that the length of the domain should not be too short or too long. The former alters turbulence characteristics while the latter can result in relaminarization. These observations were made for a turbulent oscillatory flow inside a pipe. They may not be directly applicable to oscillatory flow across parallel-plate structures, but these ideas need to be borne in mind.

Experimental work reported by Shi et al. [16,17] has shown that the introduction of a temperature distribution along the fins (compared to isothermal situation) results in flow asymmetry within the channels and in the immediate vicinity outside and temperature-driven buoyant flow, giving rise to convective currents some distance away from the plates. The flow asymmetry observed through the velocity profiles shown in [17] may be related to the temperature-dependent properties of the fluid and additional flow forcing due to natural convection.

Piccolo [12] performed numerical studies of the heat exchanger arrangement of Shi et al. [17] and further comparisons between experiment and CFD were made in [18]. The laminar model was limited only to the area between the heat exchanger plates and neglected the gravity effects. The natural convection observed in the experiment was modelled as "heat loss" by introducing a fictitious heat sink and heat source next to the heat exchangers. This approach provides a general idea about the magnitude of heat losses occurring in the experiment. However, detailed investigation needs to be conducted to identify the mechanism that contributes to these losses. Besides, buoyancy forces may also contribute to the nonlinearity of flow in addition to turbulence and streaming [8].

The current understanding of thermoacoustic processes is based on the assumption that the linear acoustic theory developed by Rott [19] is reliable for the flow with the drive ratio (the ratio of oscillating pressure amplitude to mean pressure) of up to 2–3% [1]. However, turbulence may occur at lower drive ratios particularly within the internal structures of thermoacoustic devices, and hence, numerical modelling needs to be improved. There are quite a number of turbulence models available for solving fluid dynamical problems [20]. Unfortunately, there are no strict guidelines as to which model is best for a given case. Modelling turbulence, among others, requires knowledge about the flow physics and the established practice for a specific problem. The correct choice of turbulence model is important to ensure that the physics of flow within practical thermoacoustic systems is properly presented. The objective of this study is to test the performance of widely-available RANS (Reynolds-Averaged Navier–Stokes) turbulence models for the prediction of flow behaviour in oscillatory flow conditions in the presence of the wall temperature gradient, which is a typical situation in heat exchangers of thermoacoustic devices. In addition, studies of viscous dissipation, related to pressure losses in real systems, are carried out, together with surface heat flux calculations to provide additional validation against available experimental results.

2. Computational Model

The computational domain used in this paper is based on the experimental setup of Shi et al. [16,17], sketched at the top of Figure 1. It is a loudspeaker-driven, standing-wave, 8.3 m-long resonator (of which 7.4 m is a constant cross-section square duct of 134 mm × 134 mm). It is filled with nitrogen at atmospheric pressure and operates in the quarter-wavelength mode, with the frequency of 13.1 Hz. The pressure antinode is at the end-wall. The resonator contains a parallel-plate hot heat exchanger (HHX) coupled with a cold heat exchanger (CHX). Overall, there are 10 pairs of individual plates: five pairs are heated/cooled, and five pairs (at the top/bottom) provide a uniform porosity (flow resistance) across the cross-section. The dimensions of the plates and flow channels are shown at the bottom of Figure 1. Point 'm' denotes the "joint" where the hot and cold plates meet. It is at 0.17λ from the pressure antinode. The electrical heating elements in HHX and cooling water flow in CHX aim to provide flat temperature profiles over the fins: 200 °C and 30 °C, respectively; but in practice, the heat leaks produce a slightly distorted temperature distribution as reported in [17,21]. Particle image velocimetry (PIV) and planar laser-induced fluorescence (PLIF) techniques provide the oscillatory velocity and temperature field information around the heat exchangers.

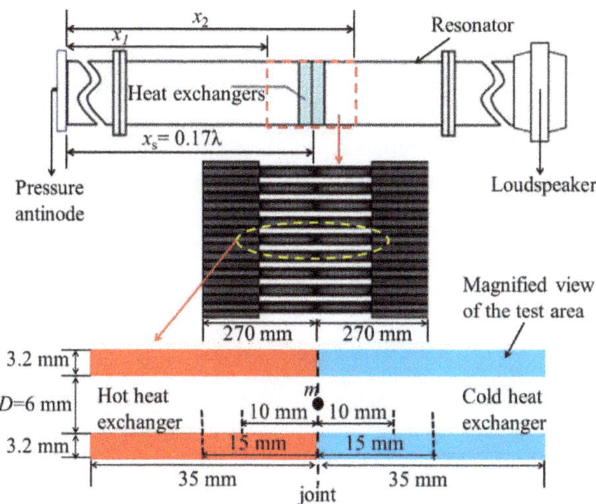

Figure 1. A "short model" shown as a red-dashed box on the schematic diagram (top); computational domain (middle); and the area of the parallel-plate channel investigated (bottom).

Preliminary numerical studies [22] have shown that the pressure and velocity in the flow far away from the heat exchangers can be estimated from the linear thermoacoustic theory. Thus, a "short model" (excluding the rest of the resonator) is acceptable if the incoming/outgoing oscillatory flow does not interfere with the plate structure [11]. The computational domain consists of two regions: the area within the plate structure and two areas outside (i.e., the open areas on the left and right from the plate structure) bounded by the resonator walls. Based on Fieldman and Wagner's correlation [13], the ratio of length to diameter (L/D) for the case investigated here should not be larger than 11.4 for flow in the channel and less than 0.52 for the flow outside. The computational domain developed based on the length of 270 mm either way from the joint between HHX and CHX (cf. the middle of Figure 1) complies with these requirements. Due to flow asymmetry caused by the natural convection observed in the experiment, the full height covering all 10 pairs of plates is covered by the model.

A quadrilateral structured mesh (cf. the middle of Figure 1) was used. It was refined in both the x- and y-directions to check for grid independency. The model was tested for cell counts of 47,360, 58,510, 71,460 and 85,560. It was found that the cell count of 71,460 provides a solution that is independent of the grid size and hence used throughout the current study. The minimum orthogonal quality is 0.86 with the maximum skewness of 0.13, which show that the mesh quality is very close to a good quality mesh (close to one for minimum orthogonality and close to zero for maximum skewness) [23]. The nearest node from the wall, $y+$, for all cases is between 0.1 and 1.24. This value differs depending on the drive ratio and flow amplitude during each phase of the flow cycle; and serves as a reference for the resolution of the mesh near the wall.

The models were solved using a high performance computer (4 login nodes, with 28 cores of 56 threads and 256 GB memory per one login node). The calculation for one cycle takes about 30 min. The transient models were calculated until a steady oscillatory condition is achieved, i.e., where pressure, velocity and temperature do not change with time. It was found that all models needed to be calculated for at least 70 cycles to meet this condition (approximately 35 h of computational time). The flow solutions presented in this paper rely on the RANS equations used for turbulence modelling, while the laminar model (see e.g., [21]) is solved for comparisons. The four RANS models tested included: k-ε, k-ω and shear-stress-transport (SST)-k-ω and transition-SST provided by ANSYS FLUENT 13.0 [23]. The RANS equations (shown in index notation) are:

$$\frac{\partial \rho}{\partial t} + \frac{\partial}{\partial x_j}(\rho u_i) = 0, \tag{1}$$

$$\frac{\partial(\rho u_i)}{\partial t} + \frac{\partial}{\partial x_j}(\rho u_i u_j) = -\frac{\partial p}{\partial x_i} + F_i + \frac{\partial}{\partial x_j}[\tau_{ij}] + \frac{\partial}{\partial x_j}\left(-\rho \overline{u_i' u_j'}\right) + \frac{\partial}{\partial x_j}\left(-\rho \overline{u_i'^2}\right), \tag{2}$$

$$\frac{\partial}{\partial t}(\rho E) + \frac{\partial}{\partial x_i}[u_i(\rho E + p)] = \frac{\partial}{\partial x_j}\left((k)_{eff}\frac{\partial T}{\partial x_j} + u_i(\tau_{ij})_{eff}\right), \tag{3}$$

where

$$(\tau_{ij})_{eff} = \mu\left(\frac{\partial u_j}{\partial x_i} + \frac{\partial u_i}{\partial x_j}\right) - \frac{2}{3}\mu_{eff}\frac{\partial u_k}{\partial x_k}\delta_{ij}. \tag{4}$$

These have a form similar to the laminar model of Navier–Stokes equations, but the solution variables now represent the ensemble-averaged (or time-averaged) values. The new terms at the end of Equation (2) are the Reynolds stresses representing the turbulence effect. For compressible flow, the fluctuation of density may affect the turbulence. In this case, the transport equations can be interpreted as Favre-averaged Navier–Stokes equations, with velocities representing the mass-averaged values [20,22]. The effective stress tensor $(\tau_{ij})_{eff}$ represents the stress tensor under the influence of turbulence with effective viscosity, $\mu_{eff} = \mu + \mu_t$, defined as a sum of laminar viscosity, μ, and turbulent viscosity, μ_t. Similarly, the effective thermal conductivity, $k_{eff} = k + k_t$, is the sum of mean thermal conductivity, k, and turbulent conductivity, k_t. The turbulent thermal conductivity is calculated as, $k_t = \mu_t c_p / Pr_t$ and the turbulent Prandtl number, Pr_t, has a constant value of 0.85. The eddy viscosity, μ_t, is calculated using the turbulence model.

The Reynolds stresses are solved through additional equations provided by the turbulence model. In this study, turbulence is assumed isotropic so that the Boussinesq hypothesis is applicable to relate the Reynolds stresses to the mean velocity gradient as follows [23]:

$$-\rho\overline{u_i'u_j'} = \mu_t\left(\frac{\partial u_i}{\partial x_j} + \frac{\partial u_j}{\partial x_i}\right) - \frac{2}{3}\left(\rho k_e + \mu_t\frac{\partial u_k}{\partial x_k}\right)\delta_{ij}. \tag{5}$$

The subscripts i, j and k refer to the coordinates x, y and z, respectively. The term, $k_e = \left(u_1'^2 + u_2'^2 + u_3'^2\right)/2$, is known as the turbulent kinetic energy. The Kronecker delta ($\delta_{ij} = 1$ if $i = j$ and $\delta_{ij} = 0$ if $i \neq j$) is introduced to correctly model the normal component of the Reynolds stress [20]. The RANS models used require additional equations to solve for turbulent viscosity, μ_t, and turbulent kinetic energy, k_e, to obtain the Reynolds stresses using Equation (5). The Reynolds stresses are then used to correctly model the momentum equation for the turbulence-affected flow through Equation (2).

The boundary conditions are calculated from the lossless equation introduced by Rott [19]:

$$P_1 = P_a cos(k_w x_1) cos(2\pi f t), \tag{6}$$

$$m_2' = \frac{P_a}{a} sin(k_w x_2) cos(2\pi f t + \theta), \tag{7}$$

$$\left.\frac{\partial T}{\partial x}\right|_{x_1, x_2} = 0. \tag{8}$$

Oscillating pressure, P_1, and mass flux, m_2', are assigned at locations x_1 and x_2, respectively (cf. Figure 1), far enough from the heat exchangers. Condition (8) is to ensure that when the flow reverses, its temperature is equal to that of the cells next to the boundary. The wave number, $k_w = 2\pi f/a$, is constant because the frequency, f, is constant. The terms a and P_a refer to the speed of sound and the pressure amplitude at the antinode, respectively. The phase, θ, is set to follow the standing-wave criterion where pressure and velocity are 90° out of phase. The adiabatic non-slip wall was assigned at the resonator walls. The heat exchanger plates were assigned temperature profiles based on experiments [17]. Figure 2 shows the interpolated profiles used in the current work. Nitrogen inside the resonator is modelled as an ideal gas. The viscosity, μ, and thermal conductivity, k, of the gas change with temperature, T, and the equations are given after Abramenko et al. [24] as:

$$\begin{gathered} \mu = 1.82 \times 10^{-5}\left(\frac{T}{T_0}\right)^{0.69}; \\ k = -8.147 \times 10^{-4} + 1.161 \times 10^{-4}T - 1.136 \times 10^{-7}T^2 + 1.062 \times 10^{-10}T^3 - \\ 5.406 \times 10^{-14}T^4 + 1.454 \times 10^{-17}T^5 - 1.942 \times 10^{-21}T^6 + 1.011 \times 10^{-25}T^7. \end{gathered} \tag{9}$$

The influence of gravity is modelled with gravitational acceleration set to 9.81 m/s^2.

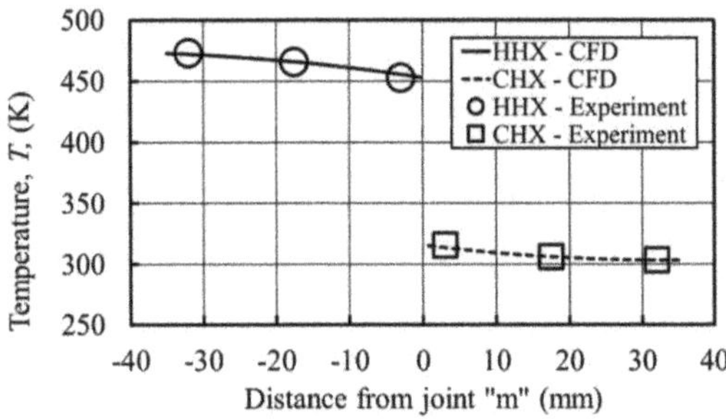

Figure 2. Temperature profiles applied to cold and hot heat exchangers based on experimental points. CHX—cold heat exchanger; HHX—hot heat exchanger; CFD—computational fluid dynamics.

Additional equations, which depend on the type of turbulence model solved, are used to solve Reynolds stresses appearing on the right-hand side of the RANS Equation (2). For all turbulence models tested, additional boundary conditions for the turbulence model are set at the boundaries x_1 and x_2 by assigning the value of turbulence intensity, I, and turbulence length scale, l_t, defined as:

$$I \equiv \frac{u'}{u_{average}} = 0.16(Re)^{-\frac{1}{8}}, \tag{10}$$

$$l_t = 0.07D. \tag{11}$$

The Reynolds number, $Re = \rho u_m D/\mu$, is calculated using the velocity amplitude at location m, u_m, obtained from Shi et al. [16,17] and the gap between plates, $D = 6$ mm, as shown in Figure 1. The density, ρ, and viscosity, μ, used in calculating Equation (10) are taken at 300 K. ANSYS FLUENT 13.0 [23] uses the inputs defined in Equations (10) and (11) to estimate the inlet distributions of turbulence kinetic energy, k_e, and turbulence dissipation, ε, using the following equations [20,22]:

$$k_e = \frac{2}{3}\left(u_{avg} I\right)^2, \tag{12}$$

$$\varepsilon = C_\mu^{3/4} \frac{k_e^{3/2}}{l_t}, \tag{13}$$

where u_{avg} is the characteristic flow velocity and C_μ is an empirical constant that varies depending on the turbulence model used (approximately 0.09). The sensitivity of the results to these inlet conditions is tested by changing the value of turbulence intensity, I, to be the value calculated using the velocity amplitude at the boundaries obtained from the equations of lossless theory. It is found that the solution is insensitive to the turbulence boundary conditions. This shows that the domain is sufficiently long to avoid the interaction between the flow around the plates and the inlet/outlet flows into/out of the domain. Default values are retained for all constants of all models used [23].

A pressure-based solver is used for all models with the application of the pressure-implicit with splitting operators (PISO) algorithm for the pressure-velocity coupling. The second order implicit discretisation scheme is selected for discretisation of time. The transport equations and turbulent equations are solved using the second order upwind scheme. This transient model is calculated with time step size set as $1/(1200f)$ per cycle, chosen so that calculations converge within 15–18 iterations in every time step. The model is set to converge at absolute values of 10^{-4} for all transport and turbulence equations and 10^{-8} for the energy equation.

3. Results and Discussion

This section presents the numerical results obtained and the relevant discussions. Section 3.1 deals with the preliminary work using the laminar model, while Section 3.2 focuses on solutions of RANS turbulence models studied in this work. Both aspects contain comparisons with earlier experimental results of Shi et al. [16,17]. The subsequent Section 3.3 discusses the numerical investigation of viscous dissipation based on the models that best represent flow behaviour in the drive ratios investigated. Finally, Section 3.4 deals with numerical heat transfer predictions and their comparisons with experimental data.

3.1. Laminar Flow Model

According to Merkli and Thomann [7], an oscillatory flow is considered laminar as long as the Reynolds number (based on the thickness of Stokes' layer) is less than 400. This is defined as $Re_\delta = 2u_m/(v\omega)^{1/2}$, with u_m, v and ω representing the axial velocity amplitude, viscosity and angular velocity, respectively. The values of the Stokes Reynolds number for all drive ratios investigated by Shi et al. [16,17] are shown in Table 1. All of these are lower than the critical value suggested in [7],

$Re_\delta = Re_c = 400$. Thus, at first, a laminar model was used to model the flow and heat transfer for the drive ratios (Dr $= P_a/P_o$; P_o is the mean pressure in the resonator) in Table 1.

Figure 3 shows the variation over the cycle of the axial velocity at the channel centre, 10 mm from the joint between the HHX and CHX plates, inside the cold channel. Results from the laminar model, for the four drive ratios, are plotted against the experimental results. For all drive ratios, the numerical and experimental values of velocity agree reasonably well, but with slight discrepancies at some phases of a flow cycle. Maximum deviation between numerical and experimental values for drive ratios of 0.30%, 0.45%, 0.65% and 0.83% are 8%, 4.7%, 10.1% and 4.6%, respectively.

Table 1. Stokes Reynolds number based on the experimental results of Shi et al. [16,17].

Drive Ratio, Dr (%)	Centreline Velocity Amplitude, u_m (m/s)	Stokes Reynolds Number, Re_{ffi}
0.30	1.30	70.25
0.45	1.90	101.49
0.65	2.97	155.34
0.83	3.84	202.33

However, the centreline velocity alone is not sufficient to represent the flow, and therefore, the instantaneous velocity profiles across the channel need to be inspected. These are plotted in Figure 4 for the same axial location, from the wall surface ($y = 0$) to the middle of the channel ($y = 3$ mm), for all 20 phases of a flow cycle, separately for each drive ratio. Clearly, the already shown mismatch between the CFD and experimental magnitudes of axial velocity at $y = 3$ mm can be seen also in Figure 4, but more importantly, the whole profiles show substantial discrepancies at other locations, especially near the wall. The over-prediction of velocity profiles near the wall as calculated by the laminar model hints at the presence of turbulence [25].

Furthermore, as the drive ratio increases, the discrepancies between velocity profiles from laminar CFD and experiments become larger, which indicates that the laminar model may not be sufficient to capture the flow behaviour. These discrepancies are also illustrated in Figure 5 as a vorticity contour map (drive ratio 0.83% selected as an example), which covers the area of the hot and cold channel. The numerical solutions appear to indicate the presence of very strong elongated secondary vortices (also referred to as a secondary shear layer), which are much less intense in reality, the best examples of such discrepancies being shown for ϕ3. Furthermore, the laminar model also produces "blobs" of vorticity that travel within the channel between phases ϕ13 and ϕ17, and this also has no match in the experimental results.

Many investigations (cf. [4,5]) have shown that oscillatory flow past parallel plate structures creates strong vortex structures at the end of plates. The remains of these structures tend to be re-entrained into the channel as the flow changes direction. This suggests that there may be sufficient instabilities present in the flow to trigger turbulent transition. Furthermore, the critical Reynolds number defined in [7] was strictly for a flow within a long, ideally infinite, channel. Therefore, it is possible that transition may be triggered at much lower Re_δ for flows past short plates. This is why a range of turbulent models has been tested as presented in the next section.

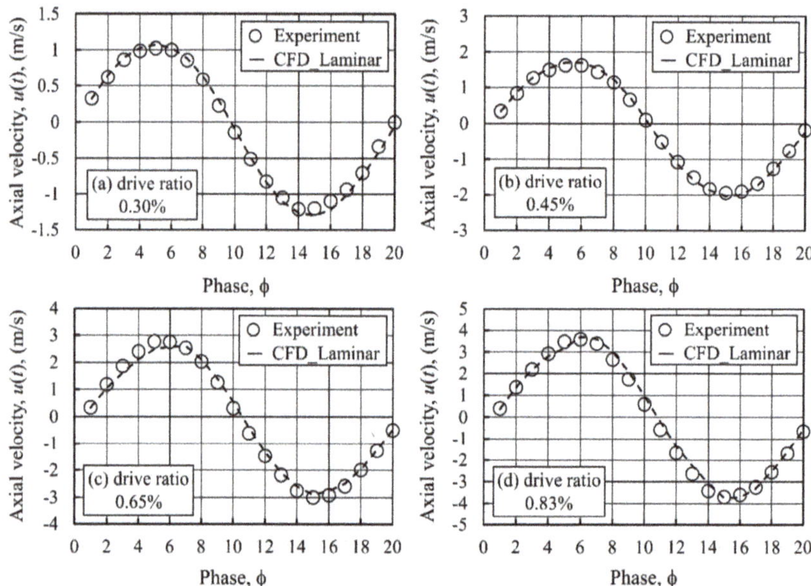

Figure 3. Axial velocity in the middle of the channel at the location 10 mm away from the joint above the cold plate for drive ratios of (**a**) 0.30%, (**b**) 0.45%, (**c**) 0.65% and (**d**) 0.83%.

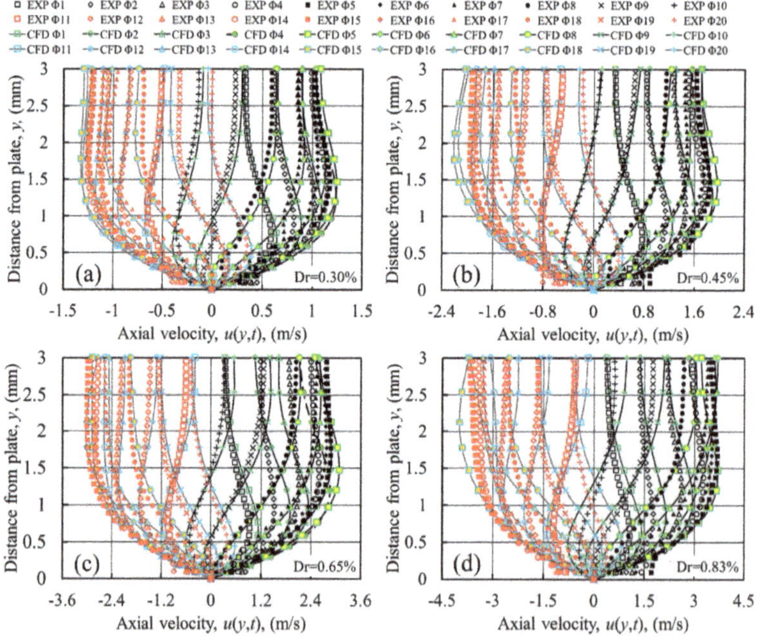

Figure 4. Velocity profiles at location 10 mm away from the joint, within the cold channel, for drive ratio (Dr): (**a**) 0.30%, (**b**) 0.45%, (**c**) 0.65% and (**d**) 0.83%. Numerical results based on laminar model. EXP refers to experimental results.

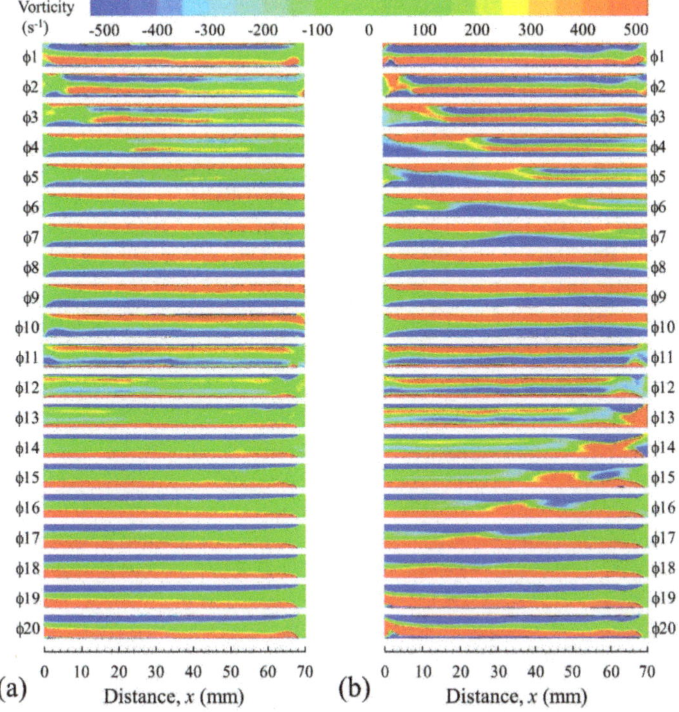

Figure 5. Comparison between vorticity contour from (**a**) experiment and (**b**) CFD laminar model. The drive ratio is 0.83%. Distance x is measured from the left end of HHX/CHX assembly; vertical axis, y is omitted for simplicity; the ratio of x:y is 1:1.

3.2. Turbulent Flow Models

Figure 6 shows the vorticity contours for drive ratio of 0.83% obtained using two turbulence models: k-ω and k-ε. These are best discussed in comparison with Figure 5. The vorticity between the plates predicted by the k-ω model exhibits weaker vortex structures compared to the results obtained from the laminar model. However, there is still a suspect "blob" of vorticity that travels within the channel that is not seen in the experiment. On the other hand, the k-ε model produces much more realistic vorticity contours. However, when comparing experimental and CFD results for phases $\phi 1$, $\phi 3$, $\phi 11$ and $\phi 13$ in Figures 5a and 6b, it is clear that the secondary shear layer (closer to the centre of the channel) is rather under-predicted by the CFD model. The results in Figures 5 and 6 suggest that the disturbances leading to the creation of the elongated vorticity regions referred to as secondary shear layers and the "blobs" of vorticity described above originate from the open area at the end of plates.

Therefore, a further investigation into the vortex structures at the end of the plates was conducted to help choose turbulence models that best describe the flow, as illustrated in Figures 7–9. Figure 7, based on experimental results at drive ratio 0.83%, shows that the flow enters the left end of the heat exchanger assembly at phase $\phi 1$ and reaches the maximum amplitude at $\phi 5$. During these phases, a weak vortex structure from the open area next to the left end of the HHX moves into the channel. The vortex structure at the right end of the heat exchanger assembly elongates as the velocity increases, but stays attached to the plate ends until the maximum velocity is reached at $\phi 5$. When the flow decelerates, the vortex structures at the right end of the plates start shedding, as seen at $\phi 8$. Similar behaviour is observed for the second half of the cycle as the flow reverses (phases ϕ11 to ϕ20), although differences in the flow patterns exist due to the asymmetry of the temperature profile (cf. Figure 2).

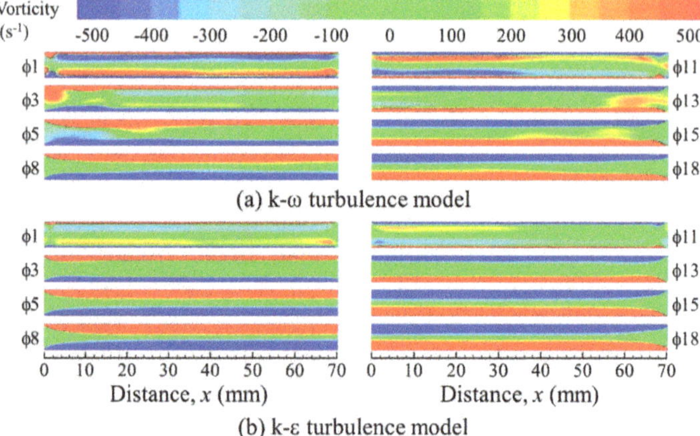

Figure 6. Comparison between vorticity contours from the (**a**) k-ω and (**b**) k-ε turbulence models. The drive ratio is 0.83%. Distance x is measured from the left end of HHX/CHX assembly; vertical axis, y, is omitted for simplicity; the ratio of x:y is 1:1.

Figure 8 shows vorticity contours at the ends of the heat exchanger assembly for drive ratio of 0.83% calculated using the k-ε model. The vortex pattern is very different form the experimental one: vortex structures at the right end of the plate for ϕ3 appear more stable. The pair of vortex structures at the end of plates elongates until the flow reaches maximum velocity at ϕ5. In the experiment, vortex structures start to "wiggle" in this phase, which the k-ε model does not seem to predict.

As the flow decelerates, the elongated structures predicted by the k-ε model retain their shapes, but start to lose their strength, which reduces considerably by the time the flow reverses at ϕ11. The weak vortex structures are re-entrained by the channel flow to create the elongated secondary vortex structures (secondary shear layers). In the second half of the cycle, similar stable vortex structures are observed on the left end of the plates. Importantly, the discrete vortex shedding phenomena are not seen in the results obtained from the k-ε turbulence model.

Figure 9 shows vorticity contours at the left and right ends of the heat exchanger assembly calculated for the drive ratio of 0.83% using the k-ω model. The vortex structures have a closer resemblance to the experimental result than those from the k-ε model. The vortex shedding process starts at the right end at ϕ1 and the left end at ϕ11. The vortex strengths seem slightly over-predicted, compared to the corresponding structures that appear somewhat weaker in the experimental results. The differences may be related to the consequence of "stitching" and "averaging" of the experimental results as explained in the original papers by Shi et al. [17] and Mao and Jaworski [4]. It may also be the result of two-dimensional assumptions used in the current turbulence model. Nevertheless, the resemblance of the flow structures to the experimental ones is quite convincing.

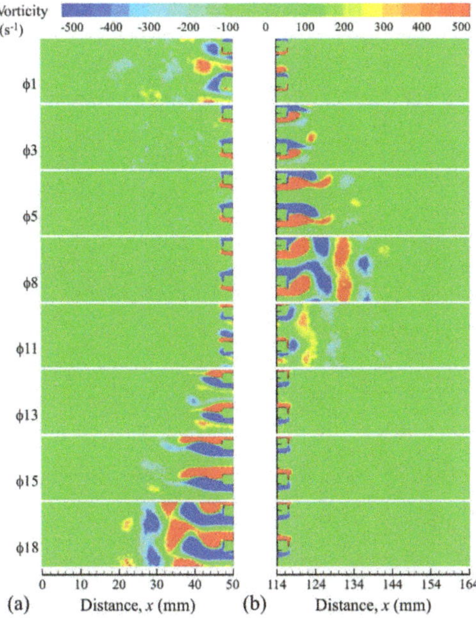

Figure 7. Vorticity contours at the left (**a**) and right (**b**) ends of the heat exchanger assembly calculated from the experimental results for the drive ratio of 0.83%. Arbitrary distance x is measured along the "viewing area"; vertical axis, y, is omitted for simplicity; the ratio of x:y is 1:1.

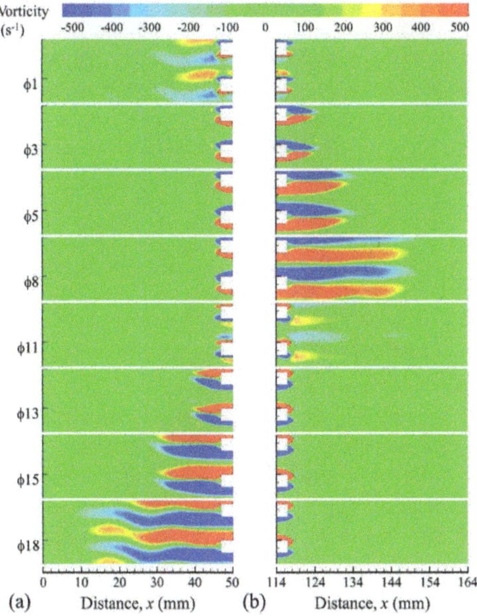

Figure 8. Vorticity contours at the left (**a**) and right (**b**) ends of the heat exchanger assembly calculated using the k-ε model for the drive ratio of 0.83%. Arbitrary distance x is measured along the "viewing area"; vertical axis, y, is omitted for simplicity; the ratio of x:y is 1:1.

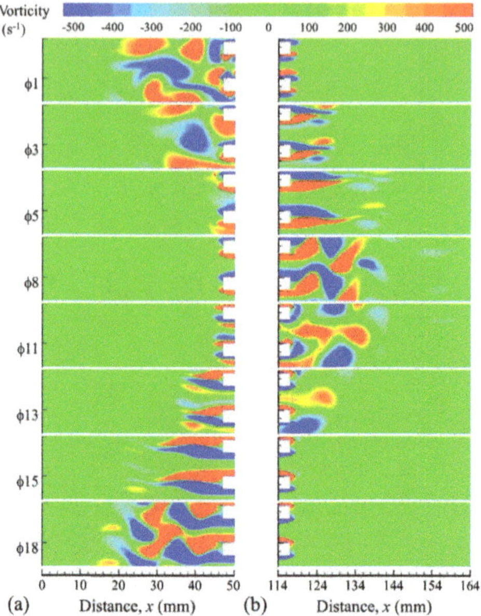

Figure 9. Vorticity contours at the left (**a**) and right (**b**) ends of the heat exchanger assembly calculated using the k-ω model for the drive ratio of 0.83%. Arbitrary distance x, is measured along the "viewing area"; vertical axis, y, is omitted for simplicity; the ratio of x:y is 1:1.

The results show that the flow structures at the end of the plates are best predicted by the k-ω turbulence model. The slight disparity with experimental results observed for the vorticity plots within the area bounded by the plates of the heat exchangers may require a more refined investigation. Menter [26] proposed a modified two-equation model called the shear-stress-transport (SST) k-ω model. It was proposed based on the observations of the performance of the conventional k-ε and k-ω models; the k-ε model is reported to perform best in the wake and free stream region while the k-ω model performs best for flows in bounded areas. Therefore, SST k-ω model uses both the k-ε and k-ω models appropriately, based on a control parameter that applies the correct model depending on the distance to the nearest wall. The model is suggested to cater for flows that involve both inviscid and viscous regions. It was used in this study to obtain better predictions and was first tested for the highest drive ratio of 0.83%.

The comparison between velocity profiles from the experimental results and the numerical models tested is shown in Figure 10 for selected phases of the flow cycle at the drive ratio of 0.83%. They are plotted at the location 10 mm away from the joint, above the cold plate. As mentioned earlier, the laminar model over-predicts the profiles especially at locations near the wall. The predictions from the k-ε and k-ω models miscalculate the magnitude of velocity both near the wall and within the flow "core" away from the wall. Taking ϕ10 as an example, the k-ε model gives good predictions at the core, but under-predicts the velocity magnitude near the wall. On the other hand, the k-ω model gives a better prediction near the wall, but under-predicts the magnitude of the velocity at the core. The profiles are better predicted by the SST k-ω model.

Figure 11 presents a comparison between experimental and numerical results from the SST k-ω model for all 20 phases. A good match, with a maximum error of 4.3%, is found for almost all phases in the flow cycle. Figure 12 gives a similar comparison in terms of vorticity contours: a very good qualitative match is found. A slight disagreement on the strength of the secondary shear layer is still visible during the first few phases that represent the acceleration stages of a flow cycle (ϕ2 to ϕ4 and ϕ12 to ϕ14). This may be related to the uncertainty of the measurement, as well as the two-dimensional assumptions of the numerical model used in a three-dimensional situation. Nevertheless, the SST k-ω model has provided the best solution for the drive ratio of 0.83% with the closest match to the experiment.

The vorticity contours at the ends of the heat exchanger assembly, calculated using the SST k-ω model, are shown in Figure 13. They agree well with the experiment and are similar to the results from the k-ω model. This indicates that the vortex structures at the end of the plates are strongly influenced by the processes in the shear layer near the wall of the channel and that a correctly-defined flow condition within the channel leads to a better prediction of flow structures at the end of the plate. Clearly, the SST k-ω model with its excellent control parameter has given an improved prediction in both areas of the flow field.

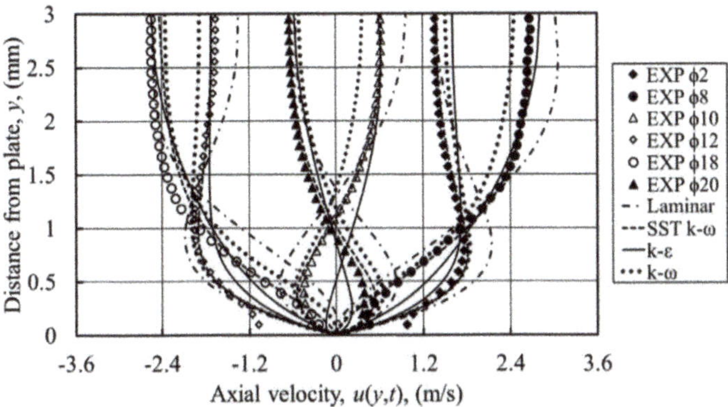

Figure 10. Comparison between velocity profiles from experiments and CFD for the location 10 mm from the joint, in the cold channel, for the drive ratio of 0.83%. SST stands for shear-stress-transport.

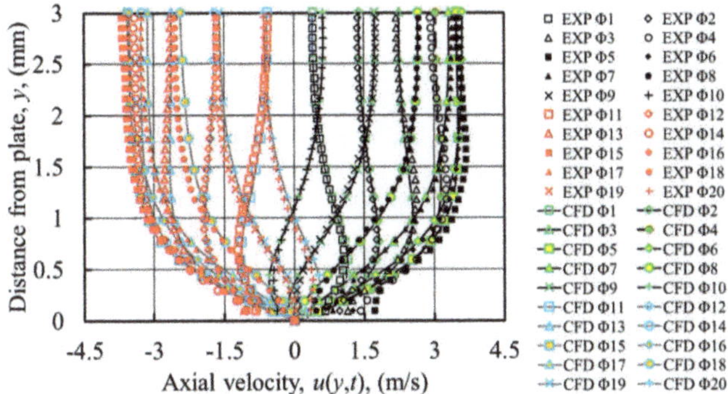

Figure 11. Velocity profiles from the SST k-ω model (black and grey lines) and experiment (red and black symbols). Plotted for the location 10 mm from the joint, inside the cold channel, for the 0.83% drive ratio.

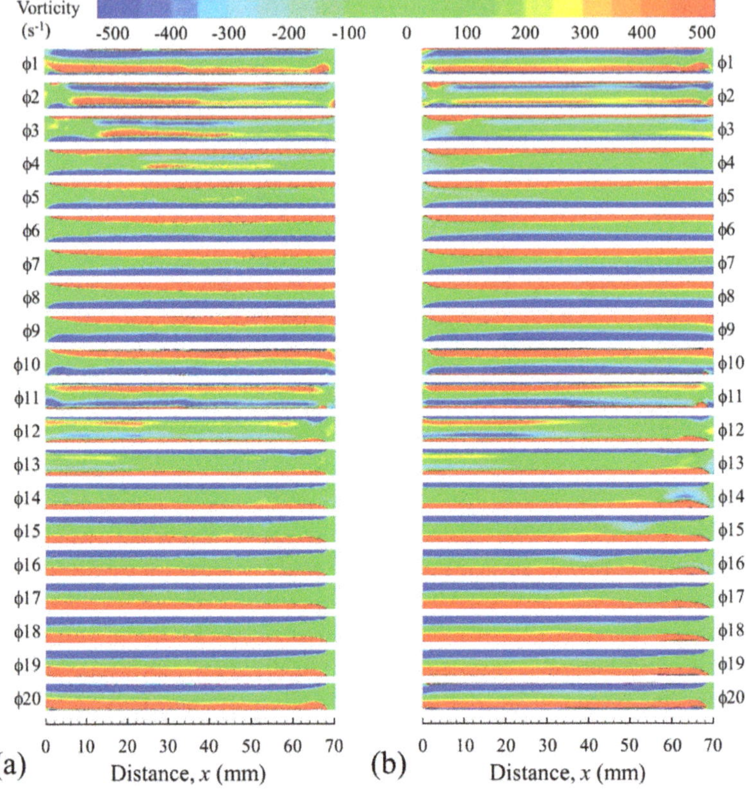

Figure 12. Comparison between vorticity contours from the experiment (**a**) and the SST k-ω model (**b**). The drive ratio is 0.83%. Distance x is measured from the left end of HHX/CHX assembly; vertical axis, y, is omitted for simplicity; the ratio of x:y is 1:1.

Following the successful application of the SST k-ω model at the drive ratio of 0.83%, it was subsequently applied to the other two drive ratios of 0.45% and 0.65%. The resulting velocity profiles for all 20 phases, together with experimental data, are shown in Figure 14. For the drive ratio of 0.65%, a better match to the experiment is obtained, compared to laminar predictions shown earlier in Figure 4c. The maximum deviation in Figure 4c occurs at ϕ5, where the velocity was numerically over-predicted by 8.9%. With the SST k-ω model, the maximum deviation is slightly reduced to 8.3%. The match is not as good as predicted for the drive ratio of 0.83%. For the drive ratio 0.45%, the SST k-ω model provides no improvement in the velocity profiles compared to the laminar model. This may indicate that the flow at the drive ratio of 0.45% has not yet reached a state suitable for the application of a turbulence model. However, the laminar model does not provide a satisfactory solution either.

Observations of the experimental and numerical results in two extreme conditions of drive ratios of 0.3% and 0.83% suggest the occurrence of transition within this range. Therefore, the four-equation transition model (transition-SST [27]) is used. Figure 15 shows the resulting velocity profiles for drive ratios of 0.45% and 0.65%. The velocity profiles for a drive ratio of 0.65% are slightly better predicted than the profile shown in Figure 14a, with the maximum error of 5.3%. Hence, the transition model seems to be a better model for predicting the flow at this drive ratio. The difference in amplitude of velocity between experimental and numerical results at ϕ5 and ϕ15 may be due to the moderation constant [21,23,26,27] not being varied in this study.

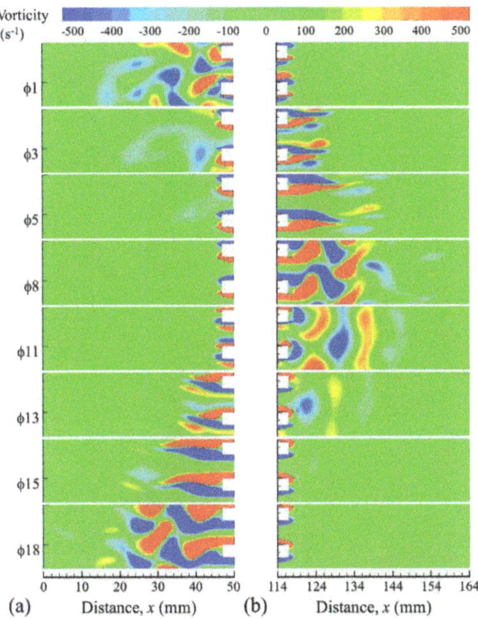

Figure 13. Vorticity contours at the left (**a**) and right (**b**) ends of heat exchanger assembly calculated from the SST k-ω model. The drive ratio is 0.83%. Arbitrary distance x is measured along the "viewing area"; vertical axis, y, is omitted for simplicity; the ratio of x:y is 1:1.

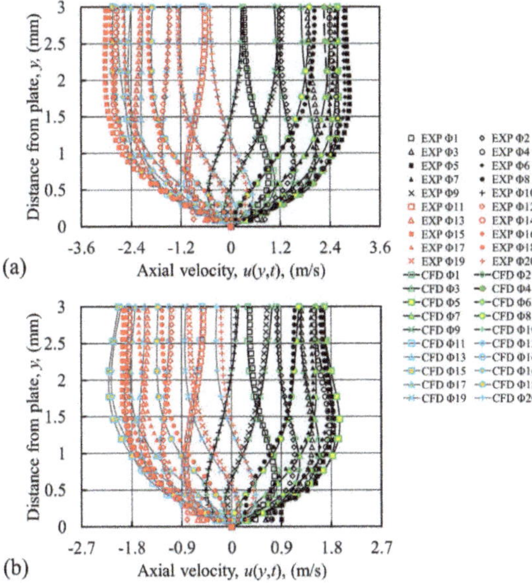

Figure 14. Comparison between velocity profiles from the experiment and the SST k-ω model; for drive ratio of (**a**) 0.65% and (**b**) 0.45%. Location 10 mm from the joint inside the cold channel.

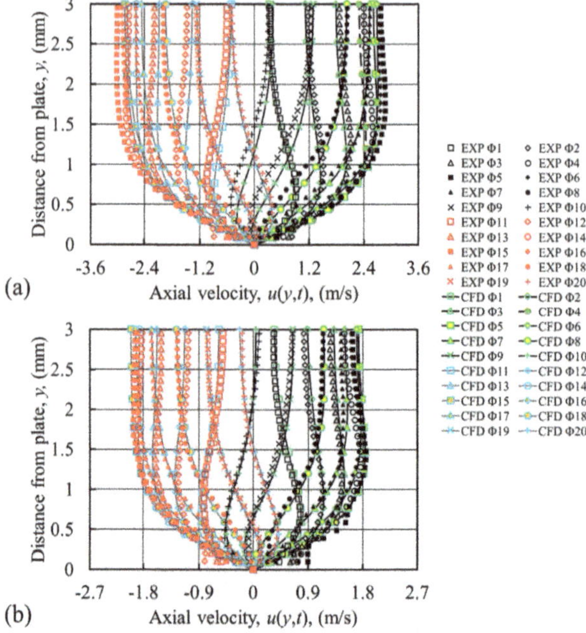

Figure 15. Comparison between velocity profiles from the experiment and the transition-SST model for drive ratio of (**a**) 0.65% and (**b**) 0.45%. Location 10 mm from the joint inside the cold channel.

For drive ratio of 0.45%, the transition-SST model provides velocity profiles that agree well with the experiment, with the maximum deviation of 5.2% (cf. Figure 15b). This may indicate that the flow at this drive ratio is already transitional in its nature. Generally, in oscillatory flow, turbulent transition is a much more complex process than in uni-directional flow. Swift [1] summarizes that the transitional region exists between the "weakly turbulent region" (flow with turbulent bursts at the centre of the pipe with the boundary layer remaining laminar) and the "conditionally turbulent region" (where the flow changes between "weakly turbulent" and "fully-turbulent" flow within a cycle). Furthermore, an oscillatory flow may also experience "relaminarization" where a "turbulent-like" flow is followed by a "laminar-like" flow in one cycle (cf. [8,9]).

As seen from Table 1, the drive ratios investigated here correspond to Stokes' layer-based Reynolds number, Re_δ, well below 400, which would be the critical Reynolds number for an infinite pipe [7]. However, the heat exchanger assembly in the current study has a short length. The disturbances from the edge of plates and the vortices shed and subsequently re-entrained (cf. Figure 13) may act as flow instabilities triggering turbulent transition. This may explain the need for the turbulence/transition model even though the Reynolds number is seemingly below the widely-accepted critical value.

As discussed above, with reference to Figure 3, the velocity amplitude at location m is not a suitable benchmark to judge the quality of CFD against experimental results. On the other hand, judging the appropriateness of the models from vorticity maps may not be entirely objective. It is possible to introduce various "metrics" to judge the quality of CFD solutions; one of the simplest resembles the concept of the standard deviation. When considering u-velocity profiles in the channel, at a selected axial location as an example, it can be defined as follows:

$$\sigma_{met} = \sqrt{\sum_{j=1}^{j=M} \sum_{i=1}^{i=N} \left[u_{exp}(y_i, \phi_j) - u_{CFD}(y_i, \phi_j)\right]^2 / MN}, \tag{14}$$

where i and j simply denote the discrete locations y between the wall and channel centreline (here, the maximum i is $N = 14$) and the phase number in the cycle (here, the maximum j is $M = 20$), respectively. It attempts to describe the deviation of the CFD-based profile from the experimental profile, or in other words, the "goodness" of predictions. The numerical values are appropriately interpolated so that y-coordinates of CFD points match with the experimental ones. Of course, the smaller the metric σ_{met}, the better the match.

Figure 16 shows the plot of metric σ_{met} for the application of three models: laminar, transition-SST and SST k-ω at the drive ratios studied. It is clear that for the smallest drive ratio, all models give a very similar value of σ_{met}. As the drive ratio increases, the laminar model gives the largest discrepancies (the metric σ_{met} grows almost linearly). Meanwhile, the transition-SST model seems to give the best predictions for drive ratios of 0.45% and 0.65%, while SST k-ω gives the best prediction for the highest drive ratio of 0.83%. This is congruent with the previous discussions that identify the transition processes for drive ratios as low as 0.45%. Of course a caveat must be made that in this example, a selected profile of velocity at the location 10 mm away from the joint above the cold plate is considered.

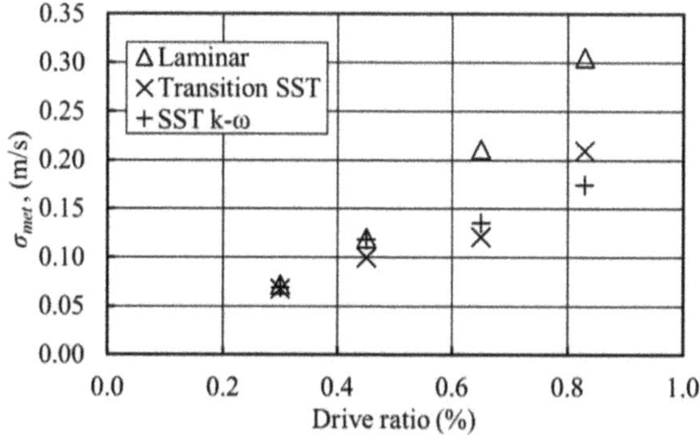

Figure 16. Plot of metric σ_{met} for laminar, transition-SST and SST k-ω models vs. the drive ratio.

The velocity magnitudes tabulated in Table 1 can also be used to represent the gas displacement, $\delta = u_m/2\pi f$; hence, the drive ratios of 0.30%, 0.45%, 0.65% and 0.83% correspond to the gas displacements of 16, 23, 36 and 47 mm, respectively. Clearly, the gas parcel in the flow with a drive ratio higher than 0.65% moves by a distance longer than half of the total length of the heat exchanger channel (70 mm). The forward and backward movements of the gas parcel convey the energy of the vortex structures at the end of plates into the channel.

Vortex structures of the selected phase, $\phi 8$, obtained from the numerical calculations for drive ratios of 0.3% and 0.83% are shown in Figure 17. Evidently, the strong vortices appearing in the plate wake for the drive ratio of 0.83% can create strong flow disturbances when pushed back into the channel. This could be a possible explanation for the appearance of turbulence (and the need for transition/turbulence model) at this drive ratio as a means of dissipating the flow energy.

The growth of any flow disturbances that occur between the plates of the heat exchanger can also be complicated by the effects of the thermal expansion that causes the gas particles to move at a different displacement amplitude for the two halves of one cycle. The short length of the plates investigated in this study prohibits the flow of high drive ratios to reach a fully-developed state. Most heat exchangers are short, and a practical system works with a high drive ratio. This study suggests that turbulence is likely to occur within heat exchangers working in the oscillatory flow condition for Stokes Reynolds numbers well below the critical value of 400 given by Merkli and Thomann [7].

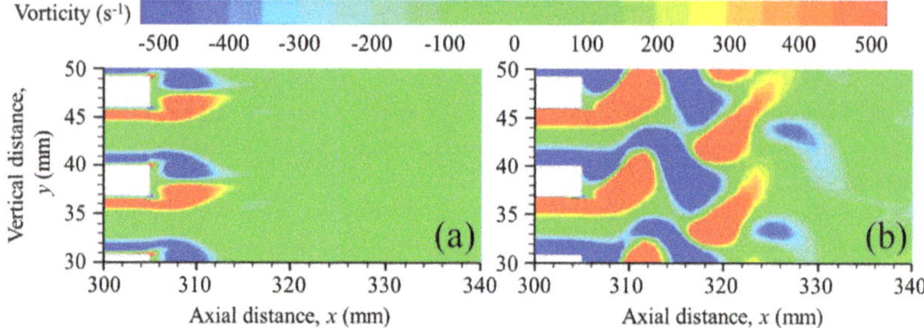

Figure 17. CFD predictions of vorticity contours at the end of the cold heat exchanger for phase ϕ8 at drive ratios of (**a**) 0.3% using the laminar model and (**b**) 0.83% using the SST k-ω model. Distances x and y measured from an arbitrary origin.

3.3. Viscous Dissipation

In a viscous flow, the energy of the fluid motion (kinetic energy) is turned into internal energy of the fluid through the existence of viscous dissipation. A dimensionless approach, introduced by Worlikar [11] and Mao [28], has been used with appropriate consideration of the porosity of the area within the heat exchanger and the open areas next to both ends of the heat exchanger (0.65 and one, respectively) [22]. The total dimensionless viscous dissipation is calculated as:

$$\langle \Phi \rangle = \sigma^2 Ec \left[\frac{1}{2\pi A} \int_0^A \int_0^{2\pi} (\mu^* \Phi^*) d\phi dA \right], \tag{15}$$

where:

$$\mu^* \Phi^* = \frac{1}{\mu_c} \left(\frac{d_c}{u_c \sigma} \right)^2 \mu \left[2 \left\{ \left(\frac{\partial u}{\partial x} \right)^2 + \left(\frac{\partial v}{\partial y} \right)^2 \right\} + \left(\frac{\partial v}{\partial x} + \frac{\partial u}{\partial y} \right)^2 - \frac{2}{3} \left(\frac{\partial u}{\partial x} + \frac{\partial v}{\partial y} \right)^2 \right]. \tag{16}$$

The terms A, μ and ϕ in Equation (15) are the area, viscosity and phase of a flow cycle, respectively; the term in square brackets is the area-weighted-average value of dimensionless dissipation, which is also averaged over one flow cycle. The dimensionless dissipation in Equation (16) is related to the mean flow velocity. The results in Figure 18 are shown as a product of porosity, σ, Eckert number, $Ec = \left(u_c^2\right) / c_{p,c} T_c$ (where u_c, $c_{p,c}$ and T_c are the reference values of velocity, specific heat and temperature, respectively) and the area-weighted-average value of dimensionless dissipation. The reference values are taken at the location 200 mm away from the end of CHX. More detailed explanations and dimensional analysis were already given by Mohd Saat and Jaworski [21]. The Eckert number, Ec, expresses the relationship between the flow kinetic energy and enthalpy and is used to characterise dissipation.

The viscous dissipation presented by Equation (16) is also known as "direct dissipation", not to be confused with turbulent dissipation [25]. The velocity components u and v in Equation (16) are the mean velocity components of the flow. As with the laminar model, the direct dissipation represents the transfer of mechanical energy to internal energy through viscosity. The turbulent dissipation (which is the transfer of energy from the mean motion into the turbulence fluctuation and then into the internal energy) is indirectly affecting the mean flow when the turbulence model is used. The effect of turbulence on the internal energy of the flow is reflected in the final velocity of the mean flow used in Equations (1), (2) and (16). The laminar model was used for the drive ratio of 0.30%, the transition-SST model for drive ratios of 0.45% and 0.65%, while the SST k-ω model for the drive ratio of 0.83%.

Figure 18 shows that the viscous dissipation increases as the drive ratio increases. The increment occurs in areas both within the channel (Figure 18a) and the area outside (Figure 18b). Hot areas have a higher viscous dissipation compared to cold ones, due to the increase of viscosity with temperature. Furthermore, thermal expansion at the area bounded by the wall of the HHX and the buoyancy-driven flow as reported in [16], particularly prominent at the hot end, may well be responsible for the higher viscous dissipation within these areas. As the drive ratio increases, the amplitude of flow increases, and the viscous dissipation becomes larger.

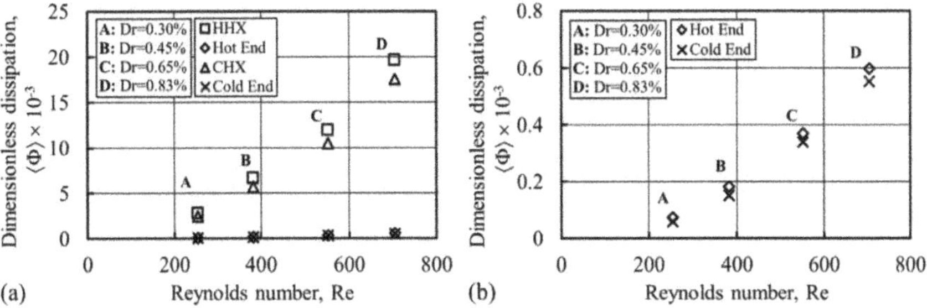

Figure 18. The effect of the drive ratio on dimensionless mean flow viscous dissipation (**a**) and the enlarged view of the viscous dissipation at the end of plates (**b**).

3.4. Heat Transfer Condition

Finally, the wall heat transfer calculated from the CFD models is compared to the experimental data. Heat flux, q, for all models is averaged over one cycle and the length of the heat exchanger, l, and calculated as:

$$q_{H,C} = \frac{1}{2\pi l}\int_0^{2\pi}\int_0^{l} -k\frac{dT}{dy}\bigg|_{wall} dx d\phi. \quad (17)$$

Subscripts H and C refer to HHX and CHX, respectively. Here, the mean thermal conductivity, k, is used to calculate the heat flux obtained from laminar, transition-SST and SST k-ω models [22].

Figure 19 shows that the heat flux predicted from the numerical models and experiments increases with an increase of the drive ratio (indeed, for easier comparisons with the literature, the drive ratio is replaced by the Reynolds number, but it is still easy to identify the four drive ratios). The Reynolds number is calculated as $Re = \rho u_m D/\mu$ as already outlined in Section 2. The flow behaviour for drive ratios investigated in this paper has been shown to be well predicted using the SST k-ω turbulence model and the transition-SST model. Therefore, the results shown in Figure 19 are based on the laminar and these two turbulent models.

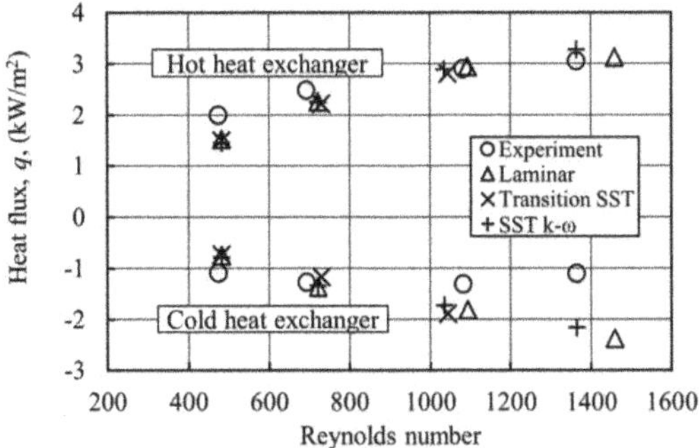

Figure 19. Heat fluxes at cold and hot heat exchangers: comparison between numerical and experimental results.

In general, judging from Figure 19, the numerical predictions of heat transfer at the HHX are reasonably well aligned with the experimental values for all drive ratios. Here, the lowest discrepancy between numerical predictions and experiment is around 24% for the lowest drive ratio of 0.30%. As the drive ratio increases, the discrepancy drops to 7%. On the other hand, for the CHX, the agreement seems reasonable for the three lower drive ratios, 0.30%, 0.45% and 0.65%, where the lowest discrepancy is around 31%, 4% and 30%, respectively. The disagreement becomes substantially worse for the highest drive ratio of 0.83%, where the smallest discrepancy between CFD and the experiment is around 97%.

Of course, a judgement as to what constitutes reasonable agreement is somewhat arbitrary; however, the literature tends to agree that in complex flow cases like this, a discrepancy of around 30% is not unexpected. One also needs to remember that the experimental results were obtained from the acetone-based PLIF measurement technique with the stated temperature accuracy of ±16 K, where calibration for the low temperature end of the range is particularly challenging. This may well explain the discrepancies between CFD and experiment that tend to be within 30% in the majority of cases, but in particular, also the case of the highest drive ratio for the CHX, which seems to be an experimental "outlier".

It is also worth comparing in detail the CFD results themselves. It is clear that the results obtained from the turbulent models are consistently closer to the experiment than from the laminar model (although this improvement is dwarfed by the overall large gap between all numerical results and the experiment). In particular, the transition-SST model gives better predictions of heat flux for drive ratios of 0.45% and 0.65% compared to predictions from the laminar model. For the highest drive ratio of 0.83%, the use of the SST k-ω, instead of the laminar, model brings the discrepancy of 117% down to 97%.

4. Conclusions

Numerical work presented in this paper builds upon previous experimental investigations [16,17] of the fluid flow and heat transfer processes in oscillatory flow conditions occurring at the coupled hot and cold plate-type heat exchangers. It presents a series of numerical studies, carried out using the ANSYS FLUENT 13.0 CFD package, aiming to match the experimental results by considering a range of flow models including laminar and four available RANS models: k-ε, k-ω, SST-k-ω and transition-SST. The key finding of this study is that in order to replicate the experimental flow fields, one would need to assume that the transition to turbulence occurs at much lower drive ratios than commonly accepted in the existing literature. In particular, the need for a turbulence model for the drive ratios as low as 0.45% has been demonstrated.

The investigation of the differences observed between the velocity profiles from the experiment and laminar model led to a hypothesis that turbulence may have influence on the flow, despite the commonly-accepted criterion for transition proposed by Merkli and Thomann [7]. This was concluded to be the effect of the finite length of the channel and the presence of the flow disturbances re-entrained after the vortex shedding cycle. Some validity to this description of the flow has been given by showing that the turbulent model gives a good match (with errors of $\approx 5\%$) between the results obtained from the experiment and the simulation, particularly at the drive ratio of 0.83% (the highest investigated). The SST k-ω model developed by Menter [26] is shown to best capture the essence of the flow at that drive ratio. The application of turbulence models makes the value of the metric σ_{met}, introduced to show the deviation between CFD and experimental velocity profiles, smaller. For instance, at the drive ratio of 0.83%, the metric value drops from 0.305 down to 0.174 as the SST k-ω replaces the laminar model.

It has been shown that transition may occur at a drive ratio as low as 0.45%, which translated into Stokes' layer-based Reynold numbers corresponds to $Re_\delta = 100$. The data shown in Table 1 would then in turn indicate the possible value of critical Reynolds number approximately within: $70 < Re_\delta < 100$. This finding is important in that it contributes to a new understanding that the turbulent flow features may appear at lower flow amplitudes within structures of thermoacoustic systems. Of course, the nature of turbulence and transition in the oscillatory flows is not the same as in relatively better understood steady flows. It is known that there are quite a few different turbulent flow regimes in oscillatory flows, and these may appear at different parts of the cycle with a possibility of relaminarization. Therefore, in the context of the current study, it may well be necessary to refine any future CFD models from the point of view of the possible turbulence regimes in the oscillatory flows.

As expected, viscous dissipation increases with drive ratio. The dissipation is also higher within the hot areas. The results give an illustration of possible heat generation within certain areas of the thermoacoustic system as a result of dissipation. The amount of energy may not be large, but it may have a significant impact on the performance of fluid flow and heat transfer within small channels, such as those within the CHX and HHX. Since dissipation is very much related to the values of velocity, the correct prediction of velocity profiles, especially near the wall, is very important. The heat exchangers are normally placed within the most important parts of thermoacoustic devices. Therefore, any contribution to the energy loses is worth considering.

The current study included the calculation of cycle- and space-averaged heat fluxes on the CHX and HHX plates. It is clear that none of the models used could replicate the experimental results very closely, and the reason for this may well be the measurement errors embedded in the experimental method. However, the general trends of heat flux versus drive ratio (or Reynolds number) appear to be correct for both HHX and CHX plates. Out of eight experimental case studied, four seem to have discrepancies below 10%, three within 31% and one within 100% (possibly an experimental "outlier"). Importantly, the application of relevant turbulence or transition models in appropriate flow conditions tends to close the gap between the CFD and experimental results.

Acknowledgments: The first author would like to acknowledge the sponsorship of the Ministry of Education, Malaysia and Universiti Teknikal Malaysia Melaka (FRGS/1/2015/TK03/FKM/03/F00274). The second author would like to acknowledge the sponsorship of the Royal Society Industry Fellowship (2012–2015), Grant Number IF110094, and the EPSRC Advanced Research Fellowship, Grant GR/T04502.

Author Contributions: Fatimah A. Z. Mohd Saat performed the numerical investigations, analysed the data and drafted the paper. Artur J. Jaworski planned and supervised the research works, outlined the paper and subsequently worked with the first author on consecutive "iterations" of the paper till the final manuscript.

Conflicts of Interest: The authors declare no conflict of interest. The founding sponsors had no role in the design of the study; in the collection, analyses or interpretation of data; in the writing of the manuscript; nor in the decision to publish the results.

References

1. Swift, G.W. *Thermoacoustics: A Unifying Perspective for Some Engines and Refrigerators*; Acoustical Society of America (through the American Institute of Physics): Melville, NY, USA, 2002.
2. Zhang, S.; Wu, Z.H.; Zhao, R.D.; Yu, G.Y.; Dai, W.; Luo, E.C. Study on a basic unit of a double-acting thermoacoustic heat engine used dish solar power. *Energy Convers. Manag.* **2014**, *85*, 718–726. [CrossRef]
3. Matveev, K.; Backhaus, S.; Swift, G. On Some Nonlinear Effects of Heat Transport in Thermal Buffer Tubes. In Proceedings of the 17th International Symposium on Nonlinear Acoustics, The Pennsylvania State University, PA, USA, 18–22 July 2006.
4. Mao, X.; Jaworski, A.J. Application of particle image velocimetry measurement techniques to study turbulence characteristics of oscillatory flows around parallel-plate structures in thermoacoustic devices. *Meas. Sci. Technol.* **2010**, *21*. [CrossRef]
5. Zhang, D.W.; He, Y.L.; Yang, W.W.; Gu, X.; Wang, Y.; Tao, W.Q. Experimental visualization and heat transfer analysis of the oscillatory flow in thermoacoustic stacks. *Exp. Therm. Fluid Sci.* **2013**, *46*, 221–231. [CrossRef]
6. Ohmi, M.; Iguchi, M.; Kakehashi, K.; Masuda, T. Transition to turbulence and velocity distribution in an oscillating pipe flow. *Bull. JSME* **1982**, *25*, 365–371. [CrossRef]
7. Merkli, P.; Thomann, H. Transition in oscillating pipe flow. *J. Fluid Mech.* **1975**, *68*, 567–575. [CrossRef]
8. Akhavan, R.; Kamm, R.D.; Shapiro, A.H. An investigation of transition to turbulence in bounded oscillatory Stokes flows Part1. Experiments. *J. Fluid Mech.* **1991**, *225*, 395–422. [CrossRef]

9. Narasimha, R.; Sreenivasan, K.R. *Relaminarization of Fluid Flows. Advances in Applied Mechanics*; Academic Press: Burlington, MA, USA, 1979; pp. 221–309.
10. Arnott, W.P.; Bass, H.E.; Respet, R. General formulation of thermoacosutic for stacks having arbitrarily shaped pore cross sections. *J. Acoust. Soc. Am.* **1991**, *90*, 3228–3237. [CrossRef]
11. Worlikar, A.S.; Knio, O.M. Numerical simulation of a thermoacoustic refrigerator I. Unsteady adiabatic flow around the stack. *J. Comput. Phys.* **1996**, *127*, 424–451. [CrossRef]
12. Piccolo, A. Numerical computation for parallel plate thermoacoustic heat exchangers in standing wave oscillatory flow. *Int. J. Heat Mass Transf.* **2011**, *54*, 4518–4530. [CrossRef]
13. Fieldman, D.; Wagner, C. On the influence of computational domain length on turbulence in oscillatory pipe flow. *Int. J. Heat Fluid Flow* **2016**, *61*, 229–244. [CrossRef]
14. Abd El-Rahman, A.I.; Abdelfattah, W.A.; Fouad, M.A. A 3D investigation of thermoacoustic fields in a square stack. *Int. J. Heat Mass Transf.* **2017**, *108*, 292–300. [CrossRef]
15. Sharify, E.M.; Takahashi, S.; Hasegawa, S. Development of a CFD model for simulation of a traveling-wave thermoacoustic engine using an impedance matching boundary condition. *Appl. Therm. Eng.* **2016**, *107*, 1026–1035. [CrossRef]
16. Shi, L.; Mao, X.; Jaworski, A.J. Application of planar laser-induced fluorescence measurement techniques to study the heat transfer characteristics of parallel-plate heat exchangers in thermoacoustic devices. *Meas. Sci. Technol.* **2010**, *21*, 115405. [CrossRef]
17. Shi, L.; Yu, Z.; Jaworski, A.J. Application of laser-based instrumentation for measurement of time-resolved temperature and velocity fields in the thermoacoustic system. *Int. J. Therm. Sci.* **2010**, *49*, 1688–1701. [CrossRef]
18. Jaworski, A.J.; Piccolo, A. Heat transfer processes in parallel-plate heat exchangers of thermoacoustic devices-numerical and experimental approaches. *Appl. Therm. Eng.* **2012**, *42*, 145–153. [CrossRef]
19. Rott, N. Thermoacoustics. *Adv. Appl. Mech.* **1980**, *20*, 135–175.
20. Versteeg, H.K.; Malalasekera, W. *An Introduction to Computational Fluid Dynamics the Finite Volume Method*, 2nd ed.; Prentice Hall: Essex, UK, 2007.
21. Mohd Saat, F.A.Z.; Jaworski, A.J. The effect of temperature field on low amplitude oscillatory flow within a parallel-plate heat exchanger in a standing wave thermoacoustic system. *Appl. Sci.* **2017**, *7*, 417. [CrossRef]
22. Mohd Saat, F.A.Z. Numerical Investigations of Fluid Flow and Heat Transfer Processes in the Internal Structures of Thermoacoustic Devices. Ph.D. Thesis, University of Leicester, Leicester, UK, 2013.
23. *ANSYS FLUENT 13.0*; User Manual, ANSYS Inc.: Canonsburg, PA, USA, 2010.
24. Abramenko, T.N.; Aleinikova, V.I.; Golovicher, L.E.; Kuz'mina, N.E. Generalization of experimental data on thermal conductivity of nitrogen, oxygen, and air at atmospheric pressure. *J. Eng. Phys. Thermophys.* **1992**, *63*, 892–897. [CrossRef]
25. Schlichting, H.; Gersten, K. *Boundary Layer Theory*, 8th revised and enlarged ed.; Springer: Berlin, Germany, 2001; pp. 76, 416–507.
26. Menter, F.R. Two-equation eddy-viscosity turbulence models for engineering application. *AIAA J.* **1994**, *32*, 1598–1605. [CrossRef]
27. Menter, F.R.; Langtry, R.; Volker, S. Transition modeling for general purpose CFD codes. *Flow Turbul. Combust.* **2006**, *77*, 277–303. [CrossRef]
28. Mao, X.; Jaworski, A.J. Oscillatory flow at the end of parallel-plate stacks: Phenomenological and similarity analysis. *Fluid Dyn. Res.* **2010**, *42*, 055504. [CrossRef]

© 2017 by the authors. Licensee MDPI, Basel, Switzerland. This article is an open access article distributed under the terms and conditions of the Creative Commons Attribution (CC BY) license (http://creativecommons.org/licenses/by/4.0/).

Article

Measurement of Heat Flow Transmitted through a Stacked-Screen Regenerator of Thermoacoustic Engine

Shu Han Hsu * **and Tetsushi Biwa**

Department of Mechanical Systems and Design, Tohoku University, Sendai 980-8579, Japan; biwa@m.tohoku.ac.jp

* Correspondence: hsu@amsd.mech.tohoku.ac.jp; Tel.: +81-022-795-3875

Academic Editor: Artur J. Jaworski

Received: 4 February 2017; Accepted: 13 March 2017; Published: 20 March 2017

Abstract: A stacked-screen regenerator is a key component in a thermoacoustic Stirling engine. Therefore, the choice of suitable mesh screens is important in the engine design. To verify the applicability of four empirical equations used in the field of thermoacoustic engines and Stirling engines, this report describes the measurements of heat flow rates transmitted through the stacked screen regenerator inserted in an experimental setup filled with pressurized Argon gas having mean pressure of 0.45 MPa. Results show that the empirical equations reproduce the measured heat flow rates to a mutually similar degree, although their derivation processes differ. Additionally, results suggest that two effective pore radii would be necessary to account for the viscous and thermal behaviors of the gas oscillating in the stacked-screen regenerators.

Keywords: regenerator; heat transfer; oscillatory flow

1. Introduction

A random stack of mesh screens is used as a regenerator of many Stirling engines, including thermoacoustic Stirling engines [1], because it achieves a high surface-area-to-volume ratio and good thermal contact with the working gas. Wide varieties of mesh numbers and wire diameters allow for flexible choices, but choosing an appropriate mesh screen is often difficult. This difficulty is attributable to tortuous flow channels inherent to the stacked-screen regenerator, which poses an obstacle from a theoretical perspective. Many experimental studies [2–7] have addressed this issue in oscillatory flow with frequencies below several tens of Hertz, aiming at gaining fundamental knowledge related to the mechanical Stirling engine. In the frequency range of thermoacoustic engines, however, experimental studies remain insufficient because they usually operate at several tens to hundreds of Hertz [1,8–13], even up to 23 kHz [14].

Our previous studies have measured the flow resistance and the acoustic power production in the regenerator [15,16] in oscillatory flow of pressurized Ar gas with a maximum frequency of 100 Hz. The results obtained with small amplitude oscillations show good agreement with predictions obtained using thermoacoustic theory [17–19] when the regenerator is assumed as a bundle of cylindrical tubes having effective radius $r_0 = \sqrt{d_h d_w}/2$ proposed by Ueda et al. [20], where d_h and d_w, respectively, denote the hydraulic diameter and the mesh wire diameter. Recently, Hasegawa et al. [21], based on results of experiments conducted at different frequencies and a fixed oscillatory velocity amplitude (=2 m/s) in air at atmospheric pressure, have also reported the applicability of r_0 for predicting the axial heat flow transported by the oscillatory gas in the regenerator. Further applicability of r_0 should be examined with different gases while varying the velocity amplitude.

Swift and Ward [22] proposed a set of equations of momentum, energy, and mass conservation for a stacked-screen regenerator by adopting the steady-flow data of Kays and London plots [23]. The equations are incorporated into the thermoacoustic calculation program—Design Environment for Low-amplitude Thermoacoustic Energy Conversion (DeltaEC) [24]. Although DeltaEC is now regarded as a standard design tool in the field of thermoacoustic engines, direct experimental verification is necessary to ascertain the heat transfer coefficient estimated from steady flow data.

This study was undertaken to test the empirical equations proposed for the axial heat transport experimentally using gas oscillations in the stacked-screen regenerator, when pressurized Ar gas is used as the working gas. Results show that the thermoacoustic theory provides a good starting point for elucidating heat transport in stacked screen regenerators, but further improvements are expected to be necessary for a more quantitative estimation.

2. Axial Heat Flow Estimated by Empirical Equations

In thermoacoustic theory [17–19], heat transport by gas oscillations along the x-axis with angular frequency ω is given by a result of hydrodynamic transport of entropy as

$$Q = \frac{1}{2}\rho_{\mathrm{m}} T_{\mathrm{m}} \int \mathrm{Re}[s\widetilde{V}] dA, \tag{1}$$

where s and V, respectively, represent the complex amplitude of entropy oscillations and axial velocity oscillations of the gas, and ρ_{m} and T_{m}, respectively, denote the temporal mean density and temperature. In addition, $\mathrm{Re}[\cdots]$ and $\widetilde{\cdots}$, respectively, show the real part and conjugate of a complex number. The surface integration is done over the cross-sectional area of the flow channel. The complex amplitude s in Equation (1) is expressed using complex amplitudes P and T of the gas pressure and temperature as

$$s = -\frac{\beta}{\rho_{\mathrm{m}}} P + \frac{c_{\mathrm{p}}}{T_{\mathrm{m}}} T, \tag{2}$$

where β signifies the coefficient of thermal expansion and c_{p} stands for the isobaric heat capacity per unit mass. For flow channels with a regular shape, T is obtained analytically. It can be expressed in a simpler form when the gas is assumed as an ideal gas. In such a case, the heat flow is given explicitly as

$$Q = -\frac{1}{2} A \mathrm{Re}\left[g P \widetilde{\langle V \rangle} \right] + \frac{1}{2} A \frac{\rho_{\mathrm{m}} c_{\mathrm{p}}}{\omega} \mathrm{Im}[g_{\mathrm{D}}] \frac{dT_{\mathrm{m}}}{dx} |\langle V \rangle|^2 \tag{3}$$

$$\text{with} \quad g = \frac{\chi_\alpha - \widetilde{\chi_v}}{(1+\sigma)(1-\widetilde{\chi_v})} \quad \text{and} \quad g_{\mathrm{D}} = \frac{\chi_\alpha + \sigma \widetilde{\chi_v}}{(1-\sigma^2)|1-\chi_v|^2},$$

where σ, ω, and A, respectively, denote the Prandtl number of the gas, angular frequency of the oscillation, and the cross-sectional area of the flow channel. In the equation, $\langle \cdots \rangle$ denote the cross-sectional average. Furthermore, $|\cdots|$ and $\mathrm{Im}[\cdots]$ signify taking the absolute value and imaginary part of a complex number; χ_j $(j = v, \alpha)$ is the thermoacoustic function associated with the thermal diffusivity α and kinematic viscosity v of the working gas. For a cylinder with radius r_0, χ_j is given as

$$\chi_j = \frac{2 J_1 \left[(i-1) \frac{r_0}{\delta_j} \right]}{(i-1) \frac{r_0}{\delta_j} J_0 \left[(i-1) \frac{r_0}{\delta_j} \right]}, \tag{4}$$

where J_1 and J_0, respectively, represent the first-order and zeroth-order Bessel functions of the first kind, i signifies the imaginary unit, and δ_j $(= \sqrt{2j/\omega})$ denotes the thermal or viscous penetration depth. We test the applicability of effective radius $r_0 = \sqrt{d_{\mathrm{h}} d_{\mathrm{w}}}/2$ by inserting it into the equation above. Moreover, we compare Q obtained by Equation (3) with the experimental heat flow that exists in the stacked-screen regenerator.

Swift and Ward [22] adopted the heat transfer coefficient of the steady flow and assumed the relation of $\langle T\rangle = \langle VT\rangle / \langle V\rangle$ to obtain the cross-sectional average $\langle T\rangle$ in the stacked-screen regenerator as

$$\langle T\rangle_{\mathrm{S}} = \frac{\beta T_{\mathrm{m}}}{\rho_{\mathrm{m}} c_{\mathrm{p}}} YP - \frac{1}{i\omega}\frac{dT_{\mathrm{m}}}{dx} G\langle V\rangle, \tag{5}$$

with

$$\begin{aligned}
&Y = \frac{\varepsilon_{\mathrm{s}} + (g_c + e^{2i\theta_P} g_v)\varepsilon_{\mathrm{h}}}{1+\varepsilon_{\mathrm{s}} + (g_c + e^{2i\theta_T} g_v)\varepsilon_{\mathrm{h}}}, G = \frac{\varepsilon_{\mathrm{s}} + (g_c - g_v)\varepsilon_{\mathrm{h}}}{1+\varepsilon_{\mathrm{s}} + (g_c + e^{2i\theta_T} g_v)\varepsilon_{\mathrm{h}}},\\
&\varepsilon_{\mathrm{s}} = \frac{\phi\rho_{\mathrm{m}} c_{\mathrm{p}}}{(1-\phi)\rho_{\mathrm{s}} c_{\mathrm{s}}}, \varepsilon_{\mathrm{h}} = \frac{i}{2b\sigma^{1/3}}\left(\frac{d_{\mathrm{h}}}{\delta_\alpha}\right)^2,\\
&b = 3.81 - 11.29\phi + 9.47\phi^2,\\
&\theta_P = \mathrm{phase}\left[\langle V\rangle\right] - \mathrm{phase}\left[P\right], \theta_T = \mathrm{phase}\left[\langle V\rangle\right] - \mathrm{phase}\left[\langle T\rangle\right],\\
&g_c = \frac{2}{\pi}\int_0^{\pi/2}\frac{dz}{1+Re_{\mathrm{h}}^{3/5}\cos^{3/5}(z)}, g_v = -\frac{2}{\pi}\int_0^{\pi/2}\frac{\cos 2z dz}{1+Re_{\mathrm{h}}^{3/5}\cos^{3/5}(z)},
\end{aligned} \tag{6}$$

where ϕ represents the volume porosity, and where both the solid density ρ_{s} and heat capacity c_{s} are safely assumed as much larger than those of the gas; Re_{h} denotes the Reynolds number given as $Re_{\mathrm{h}} = |\langle V\rangle| d_{\mathrm{h}}/\nu$ ($|\langle V\rangle|$ represents the magnitude of the complex velocity $\langle V\rangle$). The heat flow in Equation (1) is then given for an ideal gas as

$$Q \cong \frac{1}{2} A\rho_{\mathrm{m}} T_{\mathrm{m}} \mathrm{Re}\left[\langle s\rangle \widetilde{\langle V\rangle}\right], \tag{7}$$

with

$$\langle s\rangle = -\frac{1}{\rho_{\mathrm{m}} T_{\mathrm{m}}} P + \frac{c_{\mathrm{p}}}{T_{\mathrm{m}}}\langle T\rangle_{\mathrm{S}}. \tag{8}$$

We also compare Q given by Equation (7) with the value obtained from experimentation.

In contrast to the Swift–Ward formulation, Tanaka et al. and Gedeon and Wood experimentally obtained the Nusselt number Nu for oscillatory flows in the frequency range of the mechanical Stirling engine. Tanaka et al. obtained the Nusselt number from measurements conducted with a maximum oscillation frequency of 10 Hz [3], which is expressed as

$$Nu_{\mathrm{T}} = 0.33 \mathrm{Re_h}^{0.67}. \tag{9}$$

Gedeon and Wood also proposed their Nu-Re_{h} correlation [4] with oscillation frequency up to 120 Hz. They incorporated it into a Stirling machine design software code—Sage [25], which is given as

$$Nu_{\mathrm{G}} = \left(1 + 0.99\left(Re_{\mathrm{h}}\sigma\right)^{0.66}\right)\phi^{1.79}. \tag{10}$$

Applicability of these empirical equations should be tested, but incorporating them into the thermoacoustic theory presents a difficult problem for several reasons [22]. Therefore, as a tentative method, we determine $\langle T\rangle$ by inserting these empirical Nusselt numbers into the following equation:

$$\langle T\rangle_{Nu} = \left(\rho_{\mathrm{m}} c_{\mathrm{p}} + \frac{16}{i\omega}\frac{kNu}{d_{\mathrm{h}}^2}\right)^{-1} P - \left(i\omega + \frac{16}{\rho_{\mathrm{m}} c_{\mathrm{p}}}\frac{kNu}{d_{\mathrm{h}}^2}\right)^{-1}\frac{dT_{\mathrm{m}}}{dx}\langle V\rangle. \tag{11}$$

Equation (11) is obtained from Equation (27) of Reference [22] by introducing complex notation for $\langle T\rangle$ and by replacing the heat transfer coefficient h with $h = 4kNu/d_h$ while using thermal conductivity k of the gas. The heat flow can be estimated by inserting $\langle T\rangle_{Nu}$ into Equation (8) and by then using

Equation (7). Although this derivation would not qualify as a legitimate procedure, as we show later, it yields heat flows adequately to a similar degree to those of other methods tested in this study.

3. Experiments

Figure 1 presents the experimental setup, which consists of a stainless steel cylindrical tube with radius $R = 20$ mm and eight loudspeakers (FW168N; Fostex Co., Tokyo, Japan). The tube was filled with Ar gas of mean pressure of 0.45 MPa, which is one of the frequently used working fluids of the thermoacoustic engine. It contains a 35-mm-long regenerator and two 20-mm-long heat exchangers with temperatures T_H and T_R (=293 K). Regenerators of two kinds were employed: randomly stacked stainless-steel woven wire mesh screens (with mesh number of #20, #30, #50, #60, and #80) and a cylindrical ceramic honeycomb catalyst. Table 1 presents geometrical parameters of the wire meshes and ceramic honeycomb used for this study. The hydraulic diameter of the stacked-screen regenerator described in Table 1 was calculated as $d_h = \phi d_w / (1 - \phi)$. The cylindrical ceramic honeycomb catalyst possesses regular flow channels made of square pores of $d_h \times d_h$. Table 1 also includes parameters of the ceramic honeycomb catalyst.

Table 1. Geometrical properties of regenerators.

Stacked-Screen Regenerator			
Mesh Number	**Wire Diameter (mm)**	**Hydraulic Diameter (mm)**	**Volume Porosity**
#20	0.2	1.30	0.87
#30	0.22	0.80	0.78
#50	0.14	0.47	0.77
#60	0.12	0.40	0.77
#80	0.12	0.28	0.70
Ceramic Honeycomb Catalyst			
Cell Number		**Hydraulic Diameter (mm)**	**Volume Porosity**
1200		0.68	0.89

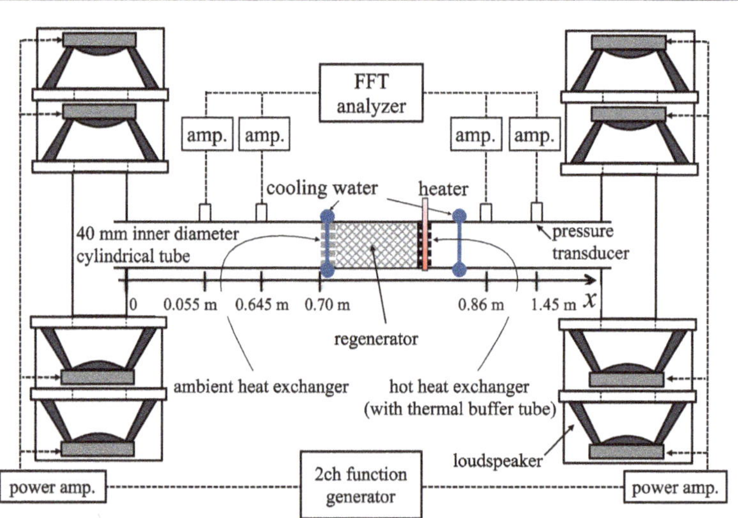

Figure 1. Schematic diagram of the experimental setup.

The hot heat exchanger with temperature T_H was heated using three electrical heater rods, whereas the ambient heat exchanger with T_R was cooled by running water around it. Temperatures T_H and T_R

were monitored using Type-K thermocouples placed on the central axis and at the regenerator ends, from which we determined the axial temperature difference $\Delta T = T_{\rm H} - T_{\rm R}$ across the regenerator. One end of a 50-mm-long thermal buffer tube region, next to the hot heat exchanger, was also cooled by running water.

The experimental procedure is explained as follows. First, in the absence of acoustic oscillations, a steady non-oscillating state with ΔT = 250 K was established. The necessary heating power was recorded as $Q_{\rm OFF}$, which turned out to be 75 W with the change of $\pm$4 W at most, depending on the mesh number and ambient temperature. Second, we turned on the loudspeakers at ends of the tube to excite acoustic oscillations in the setup. In the presence of acoustic oscillations, the heat power $Q_{\rm ON}$, which is necessary to maintain the steady oscillation state with ΔT = 250 K, was increased from $Q_{\rm OFF}$, which means that the oscillatory flows cause additional heat transport along the regenerator. Therefore, we determined the rate of heat flow passing through the regenerator as

$$|Q| = Q_{\rm ON} - Q_{\rm OFF}. \tag{12}$$

Measurements were repeated four times for each experimental condition to obtain the mean value and the standard deviation of the measured $|Q|$. Because the heat flow Q is directed in the negative x-direction, we discuss the heat flow rate $|Q|$ for brevity.

During the measurements, gas oscillations were measured using two pairs of pressure transducers (PD104; JTEKT Corp., Osaka, Japan) mounted on the sidewall of the cylindrical tube at the positions presented in Figure 1. From the pressure amplitudes and phases obtained with a 24 bit fast Fourier transform analyzer (DS-3100; Ono Sokki Co. Ltd., Yokohama, Japan), we evaluated the acoustic pressure field $p(x) = {\rm Re}\left[P(x)\exp\left(i\omega t\right)\right]$ and the velocity field $\langle u(x)\rangle = {\rm Re}\left[\langle U(x)\rangle\exp\left(i\omega t\right)\right]$ in the 40-mm tube using the two-sensor method [26,27]. The complex amplitude $\langle V\rangle$ of the cross-sectional average velocity in the hot end of the regenerator was derived as $\langle V\rangle = \langle U_{\rm H}\rangle / \phi$ from the continuity of volume velocity, where subscript H denotes the location at the hot end of the tube.

Throughout the experiments, the loudspeakers were controlled by voltage signals fed from a two-channel function generator via power amplifiers. By adjusting the phase difference and amplitude of the driving voltage signals, the specific acoustic impedance, given by the ratio $P_{\rm H}/\langle U_{\rm H}\rangle$ of complex amplitudes of pressure and velocity, was tuned to have magnitude of $0.3\rho_{\rm m}a \pm 20\%$ and phase angle of $0^\circ \pm 30^\circ$ at the position of the hot end of the regenerator. It should be noted that the efficient thermoacoustic Stirling engines achieved the specific acoustic impedance as high as $30\rho_{\rm m}a$ in the regenerator region [1]. In the present experimental setup, although it was equipped with eight loudspeakers, was not capable of producing such a high acoustic impedance while maintaining the magnitude of the velocity amplitude.

This amplitude ratio simplifies the first term $Q_{\rm A}$ on the right-hand side of Equation (3), and the ratio $Q_{\rm A}$ over the second term $Q_{\rm D}$ on the right-hand side of Equation (3) is then given as

$$\frac{Q_{\rm A}}{Q_{\rm D}} = \frac{{\rm Re}\left[g\right]}{{\rm Im}\left[g_{\rm D}\right]}\frac{0.3a\omega}{c_{\rm p}\left(dT_{\rm m}/dx\right)}. \tag{13}$$

In the case of the regular flow channel regenerator made of the ceramic honeycomb catalyst, the amplitude ratio $Q_{\rm A}/Q_{\rm D}$ was always below 0.09 for the present experimental conditions. For this reason, we can expect that the heat flow Q is well approximated by $Q_{\rm D}$, and therefore a quadratic function of $|\langle V\rangle|^2$. For reference, we evaluated the ratio $Q_{\rm A}/Q_{\rm D}$ for stacked screen regenerators by inserting Ueda's effective radius r_0 into Equation (3). Results showed that $Q_{\rm A}/Q_{\rm D} < 0.1$ for all mesh screens employed. Therefore, if the result of the thermoacoustic theory is also applicable to the stacked-screen regenerator, then Q would also be approximated as $Q_{\rm D}$. We use this result to elucidate the function ${\rm Im}\left[g_{\rm D}\right]$ experimentally, after the comparison of Q determined using the empirical equations and using experimentation.

4. Results and Discussion

4.1. Regular Flow Channel Regenerator

Figure 2 presents relations between the heat flow rate $|Q|$ and the axial mean oscillation velocity amplitude $|\langle V \rangle|$ at the hot end of the regenerator region for the ceramic honeycomb catalyst. The experimental $|Q|$ values were obtained using oscillation frequencies of 5 Hz and 200 Hz. Results show that the experimental $|Q|$ increases quadratically with $|\langle V \rangle|$, as expected from Equation (3). For comparison, we evaluated heat flow rate $|Q|$ and $|Q_D|$ as a function of $|\langle V \rangle|$ by inserting the relation $P = 0.3\rho_m a\phi |\langle V \rangle|$, the experimental angular frequency ω and thermal properties of the gas at hot end temperature T_H into Equation (3). The temperature gradient, dT_m/dx, was approximated with the ratio of temperature difference ΔT over the regenerator length (=35 mm), and also half of the hydrodynamic diameter, $d_h/2 = 0.34$ mm, was used as the channel radius. The cross-sectional area A was determined as $A = \phi \pi R^2$, where R is the tube radius. Results show that $|Q|$ and $|Q_D|$ are mutually close, as described in the preceding section. Furthermore, good agreement is observed between the theoretical heat flows and the measured ones. These results confirm the validity of the thermoacoustic theory. It is noteworthy that the agreement between theoretical and experimental $|Q|$ is still good, even when $|\langle V \rangle| \approx 1$ m/s with 5 Hz. The corresponding displacement amplitude becomes as large as 31 mm, which is comparable to the regenerator length (=35 mm). This result indicates that the thermoacoustic theory that assumes a uniform channel is still useful for such a large amplitude regime.

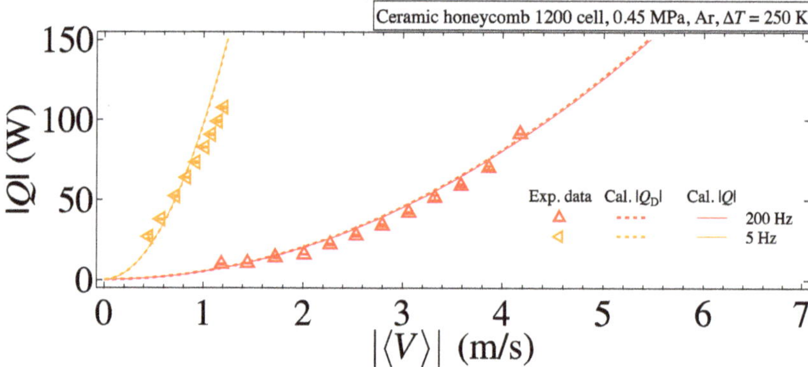

Figure 2. Comparison between the experimental heat flow rate $|Q|$ and theoretical calculations is given by Equation (3) for the ceramic honeycomb catalyst, where experimentally obtained results are shown as symbols with error bars. The theoretical calculations of $|Q|$ and $|Q_D|$ are shown, respectively, as solid and dashed curves.

4.2. Stacked-Screen Regenerator

Figure 3 presents the relation between the heat flow rate $|Q|$ and the velocity amplitude $|\langle V \rangle|$ for the mesh screens in Table 1, for 200 Hz frequency of the oscillatory flow. For comparison with the empirical equations, we show the measured $|Q|$ values in the four graphs Figure 3a–d. Solid symbols, each of which corresponds to meshes listed in Table 1, show that $|Q|$ increases quadratically with $|\langle V \rangle|$, as in the case of the regular flow channel regenerator.

Figure 3a compares the measured $|Q|$ with that obtained by inserting Ueda's effective radius r_0 into Equations (3) and (4), which shows good agreement with #50, #60, and #80 meshes, but slight deviation is visible for #20 and #30 meshes. Although the effective radius r_0 proposed by Ueda et al. was obtained from measurements at uniform temperature, the present experimentally obtained results confirm the usefulness of evaluating the heat flow in the stacked-screen regenerator when Ar is the working gas.

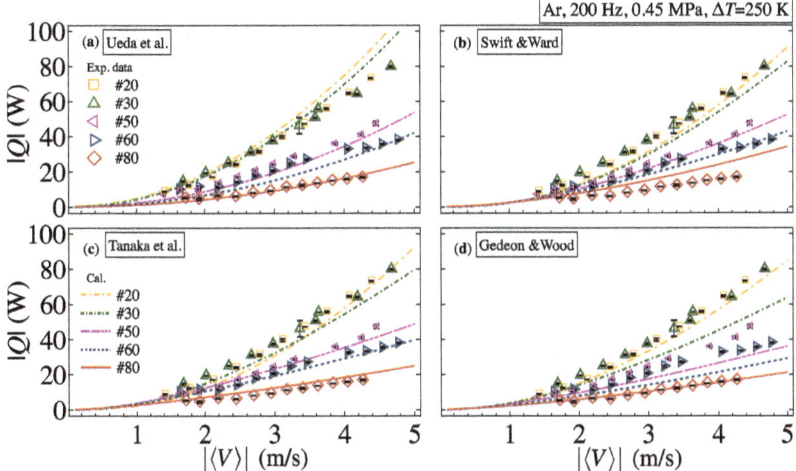

Figure 3. Experimental heat flow rate $|Q|$ and predictions given by respective empirical equations cited in this study for stacked-screen regenerators tested in pressurized helium gas with 0.45 MPa and with oscillation frequency of 200 Hz. Each panel presents identical experimentally obtained results by symbols with error bars. The empirical predictions are given, respectively, as (**a**) for Ueda et al.; (**b**) for Swift and Ward; (**c**) for Tanaka et al.; and (**d**) for Gedeon and Wood.

Comparison with results of the simple harmonic analysis derived by Swift and Ward is presented in Figure 3b, where curves are shown using Equations (5)–(8). For Equations (6), θ_T was estimated from Equation (4.68) of Reference [17] using $d_h/2$. Figure 3b shows that good agreement is obtained, except for the #80 mesh. Because the frequency of 200 Hz gives the thermal penetration depth $\delta_\alpha = \sqrt{2\alpha/\omega} = 0.15$ mm, these results indicate that the empirical expression of Swift and Ward is suitable for derivation of $|Q|$, when $d_h/(2\delta_\alpha)$ is in the range of $d_h/(2\delta_\alpha) > 1$, even though they assumed good thermal contact between the oscillation gas and the solid wall of the flow channel [24].

Heat flows estimated using the empirical equations of Tanaka et al. and Gedeon and Wood are shown, respectively, in Figure 3c,d, where we used Equations (9) and (10) to derive the temperature $\langle T \rangle_{Nu}$ in Equation (11). Then, we used Equations (7) and (8). All curves in Figure 3c show agreement with the measured $|Q|$ to a similar degree to those portrayed in Figure 3a,b. Actually, Figure 3d shows poor agreement with meshes of #30, #50, and #60, but the estimation is not bad overall. These results indicate that these empirical equations are applicable to thermoacoustic Stirling engines that operate at higher oscillation frequencies than the mechanical Stirling engines [3,4].

To assess the frequency dependence of the heat flow Q, the regenerator stacked with #30 mesh screens was tested with frequencies of 140 Hz and 180 Hz in the manner described above. Results are presented in Figure 4a–d, where the experimental $|Q|$ values increased slightly because of the decrease of the frequency. Figure 4a shows that the agreement with the effective radius of Ueda et al. was satisfactory. The Swift–Ward formulation in Figure 4b also works well, probably because of the value $d_h/(2\delta_\alpha)$ beyond 2 with those frequencies. The empirical equation of Tanaka et al. in Figure 4c agrees with the measured $|Q|$, although only the tendency of $|Q|$ for the frequency change is reproduced by the empirical equation of Gedeon and Wood, as shown in Figure 4d.

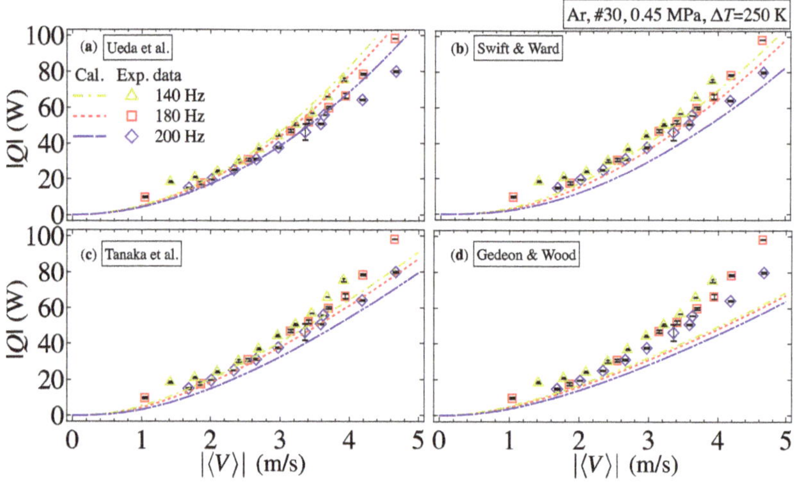

Figure 4. Comparisons between the experimental heat flow $|Q|$ and predictions given by the empirical equations. The mesh number of the stacked screens is #30. The working gas is Ar gas with 0.45 MPa. The test frequencies are 140, 180, and 200 Hz. Every panel presents the same experimentally obtained results by symbols with error bars, and empirical predictions are given respectively as the following: (**a**) for Ueda et al.; (**b**) for Swift and Ward; (**c**) for Tanaka et al.; and (**d**) for Gedeon and Wood.

4.3. Discussion

We have presented, through comparisons with experimentally obtained values, that the empirical equations proposed up to this point are useful as a first approximation to the heat flow in the stacked-screen regenerators, in spite of the considerable difference between their derivations. Among them, Ueda's formulation is the simplest to handle because of the absence of the Nusselt number Nu, which changes with the Reynolds number, Re_h, through the velocity amplitude $|V|$. Hereinafter, to strengthen the applicability of r_0 further, we investigate the function $\mathrm{Im}\,[g_D]$ from the experimental $|Q|$.

As described in Section 3, the heat flow component Q_A is negligibly small compared to Q_D for all experimental conditions. Therefore, we approximate Q with Q_D, to express $\mathrm{Im}\,[g_D]$ as

$$\mathrm{Im}\,[g_D] = \frac{2}{A}\frac{\omega}{\rho_m c_p}\frac{Q}{(dT_m/dx)\,|\langle V\rangle|^2}. \tag{14}$$

All factors in Equation (14) were evaluated using the measured values and the thermal properties of the gas at the hot end of the regenerator (T_H=543 K). The resulting $\mathrm{Im}\,[g_D]$ is shown in Figure 5, after averaging over different values of $|\langle V\rangle|$. Additionally, the error bars attached on the symbols denote the standard deviations, which show the degree of the influence of $|\langle V\rangle|$. The curve represents $\mathrm{Im}\,[g_D]$ given using $\sigma = 0.66$. It is apparent that $\mathrm{Im}\,[g_D]$ obtained by Equation (14) roughly approximates the theoretical value for the cylindrical flow channel, which means that non-dimensional quantity r_0/δ_α captures the mechanism of the oscillation-induced heat flow in the stacked-screen regenerator. This finding is consistent with the results presented by Hasegawa et al. [21], who used air at atmospheric pressure as the working gas. Therefore, we consider that the local heat transfer between the screen meshes and the working fluid would be governed mostly by the non-dimensional parameter r_0/δ_α. Therefore, the axial heat flow through the stacked screen regenerator is fundamentally described by Equation (3).

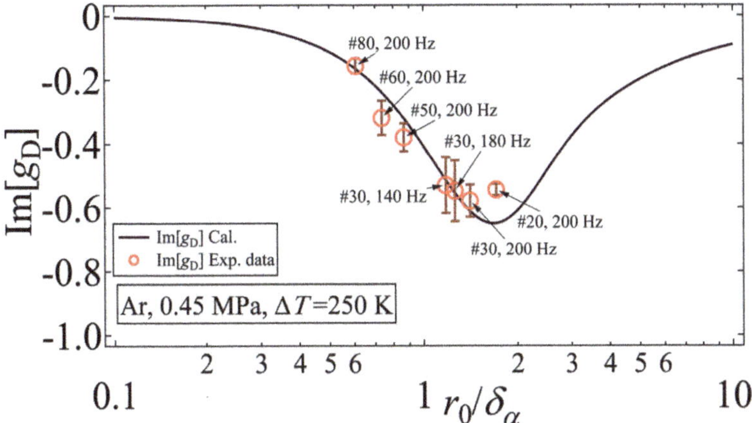

Figure 5. Relation between Im $[g_D]$ and r_0/δ_α. The curve of Im $[g_D]$ was calculated from the thermoacoustic theory. The symbols represent the experimental values evaluated from Equation (14) with measured data and the thermal properties of the gas at the hot end of the regenerator ($T_H = 543$ K).

Thermoacoustic theory describes that the flow resistance of the regular flow channel is a constant [17–19], independent of the velocity amplitude, whereas that in the porous media with tortuous flow channels is shown to increase with the velocity [28–30] or Reynolds number [2–5,7]. From measurements of the acoustic field, we have proposed modification of the effective radius r_0 to have velocity dependence [16] as

$$r_{\text{eff}} = 2\delta_\nu \left(\frac{0.8\nu}{\omega d_h^2} Re_h + \frac{16\delta_\nu^2}{d_h d_w} \right)^{-\frac{1}{2}}. \tag{15}$$

The effective radius r_{eff} approaches r_0 when the velocity is negligibly small. It decreases concomitantly with increasing velocity. Figure 6 presents a comparison of the effective radii in Equation (15) and that obtained with the experimental acoustic fields of #30 and #80 mesh screens given in Figure 4. We have determined the experimental effective radius shown by symbols in the way that the difference of the calculated acoustic powers from the measured acoustic powers at two ends of the regenerator [16] is minimized. Because the experimental effective radius reflects all mesh screens stacked in the regenerator holder, the kinematic viscosity ν in Equation (15) is evaluated at average temperature $(T_R + T_H)/2$. In addition, the horizontal axis represents the spatial average velocity amplitude $|V_m|$ over the regenerator area. The experimental effective radii were obtained with error $< 8\%$ between the measured and calculated acoustic powers. It apparently decreases with velocity $|V_m|$. Therefore, the acoustic field at the sides of the regenerator is better estimated using effective radius r_{eff} than r_0, when the velocity amplitude is finite.

Figure 7 presents comparison of the heat flow rates $|Q|$ obtained by inserting r_0 and r_{eff} into Equations (3) and (4). Results show that $|Q|$ obtained with r_{eff} yields a faulty result because it goes away from that obtained with r_0 with increasing $|\langle V \rangle|$. This result indicates that r_0 should be used instead of r_{eff} for better estimation of heat flow. Other earlier studies [29,31] have demonstrated that two effective radii are necessary to explain the viscous behavior and the thermal behavior in tortuous porous media because the viscous effects are governed by narrower regions of the channel, whereas the thermal effects are determined by wider regions [32,33]. Our experimentally obtained results provide another example representing the need of the effective thermal radius, in addition to the effective viscous radius.

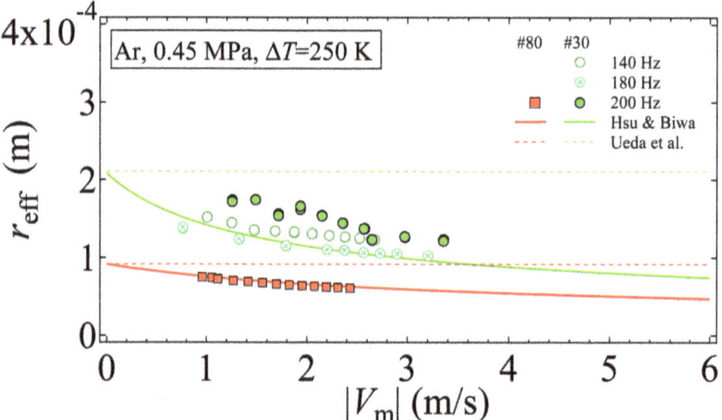

Figure 6. Relation between r_{eff} and $|V_{\mathrm{m}}|$. Symbols represent the experimental effective radius of the stacked-screen regenerator of #30 and #80 meshs. Curves stand for Equation (15) with thermal properties of the gas determined from a temperature averaged in T_{R} and T_{H}. Horizontal dashed lines represents the effective radius r_0 of Ueda et al.

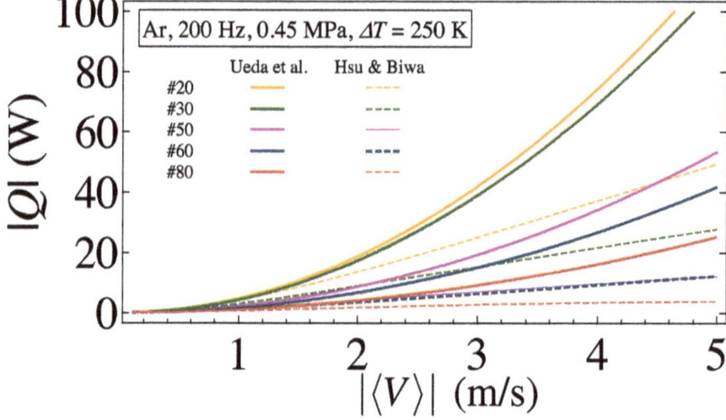

Figure 7. Comparisons between predictions of heat flow $|Q|$ given, respectively, by the effective radius of Ueda et al. and Hsu and Biwa, for stacked-screen regenerators in pressurized argon gas with 0.45 MPa and oscillation frequency of 200 Hz.

5. Conclusions

Heat transport through the stacked screen regenerators has been investigated experimentally in Ar gas with mean pressure of 0.45 MPa. The measured heat flow rates are compared with those calculated from four empirical equations, parameterized by r_0/δ_α and/or Re_{h}. Results have shown that all the empirical equations tested reproduce the measured heat flow rates to a mutually similar extent. Among those empirical equations, Ueda's formulation based on the effective radius r_0 involves the simplest derivation of the heat flow rate. Therefore, it is expected to be useful for designing thermoacoustic Stirling engines. However, the acoustic field, especially the acoustic power, can be described better by the effective radius r_{eff}, which decreases with the velocity amplitude. Therefore, it is necessary to use two effective radii to account for the viscous behavior and the thermal behavior of the gas in the stacked screen regenerators.

Acknowledgments: This work was financially supported by the Japan Science and Technology Agency (JST) through the Advanced Low Carbon Technology Research and Development Program (ALCA).

Author Contributions: Tetsushi Biwa conceived the research plan. Shu Han Hsu designed and performed the experiments and prepared the manuscript. Tetsushi Biwa revised the manuscript. Both authors have read and approved the final manuscript.

Conflicts of Interest: The authors declare no conflict of interest.

References

1. Backhaus, S.; Swift, G.W. A thermoacoustic Stirling heat engine. *Nature* **1999**, *399*, 335–338.
2. Tong, L.S.; London, A.L. Heat-transfer and flow-friction characteristics of woven-screen and cross-rod matrices. *Trans. ASME* **1957**, *10*, 1558–1570.
3. Tanaka, M.; Yamashita, I.; Chisaka, F. Flow and heat transfer characteristics of the Stirling engine regenerator in an oscillating flow. *JSME Int. J.* **1990**, *33*, 283–289.
4. Gedeon, D.; Wood, J.G. *Oscillating-Flow Regenerator Test Rig: Hardware and Theory With Derived Correlations for Screens and Felts*; NASA Contractor Report 198442; NASA-Lewis Research Center: Cleveland, OH, USA, 1996.
5. Isshiki, S.; Sakano, A.; Ushiyama, I.; Isshiki, N. Studies on flow resistance and heat transfer of regenerator wire meshes of Stirling engine in oscillatory flow. *JSME Int. J. Ser. B* **1997**, *40*, 281–289.
6. Zhao, T.S.; Cheng, P. Oscillatory heat transfer in a pipe subjected to a laminar reciprocating Flow. *Int. J. Heat Mass Transf.* **2005**, *48*, 2473–2482.
7. Costa, S.C.; Tutar, M.; Barreno, I.; Esnaola, J.A.; Barrutia, H.; García, D.; González, M.A.; Prieto, J.I. Experimental and numerical flow investigation of Stirling engine regenerator. *Energy* **2014**, *72*, 800–812.
8. Wheatley, J.; Hofler, T.; Swift, G.W.; Migliori, A. An intrinsically irreversible thermoacoustic heat engine. *J. Acoust. Soc. Am.* **1983**, *74*, 153–170.
9. Tijani, M.E.H.; Spoelstra, S. A high performance thermoacoustic engine. *J. Appl. Phys.* **2011**, *110*, 093519.
10. Wu, Z.; Man, M.; Luo, E.; Dai, W.; Zhou, Y. Experimental investigation of a 500 W traveling-wave thermoacoustic electricity generator. *Chin. Sci. Bull.* **2011**, *56*, 1975–1977.
11. Yu, Z.; Jaworski, A.J.; Backhaus, S. Travelling-wave thermoacoustic electricity generator using an ultra-compliant alternator for utilization of low-grade thermal energy. *Appl. Energy* **2012**, *99*, 135–145.
12. Smoker, J.; Nouh, M.; Aldraihem, O.; Baz, A. Energy harvesting from a standing wave thermoacoustic-piezoelectric resonator. *J. Appl. Phys.* **2012**, *111*, 104901.
13. Wang, K.; Sun, D.; Zhang, J.; Xu, Y.; Luo, K.; Zhang, N.; Zou, J.; Qiu, L. An acoustically matched traveling-wave thermoacoustic generator achieving 750 W electric power. *Energy* **2016**, *103*, 313–321.
14. Flitcroft, M.; Symko, O.G. Ultrasonic thermoacoustic energy converter. *Ultrasonics* **2013**, *53*, 672–676.
15. Obayashi, A.; Hsu, S.H.; Biwa, T. Amplitude dependence of thermoacoustic properties of stacked wire meshes. *J. Cryog. Soc. Jpn.* **2012**, *47*, 562–567. (In Japanese)
16. Hsu, S.H.; Biwa, T. Modeling of a stacked-screen regenerator in an oscillatory flow. *Jpn. J. Appl. Phys.* **2016**, *56*, 017301.
17. Swift, G.W. Thermoacoustics: A Unifying Perspective for Some Engines and Refrigerators. *J. Acoust. Soc. Am.* **2002**, doi:10.1121/1.1561492.
18. Tominaga, A. *Fundamental Thermoacoustics*; Uchida Rokakuho Publishing Co.: Tokyo, Japan, 1998; Chapter 7. (In Japanese)
19. Tominaga, A. Thermodynamic aspects of thermoacoustic theory. *Cryogenics* **1995**, *35*, 427–440.
20. Ueda, Y.; Kato, T.; Kato, C. Experimental evaluation of the acoustic properties of stacked-screen regenerators. *J. Acoust. Soc. Am.* **2009**, *125*, 780–786.
21. Hasegawa, S.; Ashigaki, Y.; Senga, M. Thermal diffusion effect of a regenerator with complex flow channels. *Appl. Therm. Eng.* **2016**, *104*, 237–242.
22. Swift, G.W.; Ward, W.C. Simple harmonic analysis of regenerators. *J. Therm. Heat Transf.* **1996**, *10*, 652–662.
23. Kays, W.M.; London, A.L. *Compact Heat Exchangers*; McGraw-Hill: New York, NY, USA, 1964.
24. Ward, B.; Clark, J.; Swift, G.W. *Design Environment for Low-Amplitude ThermoAcoustic Energy Conversion (DeltaEC) User's Guide (Version 6.3)*; Los Alamos National Laboratory: Los Alamos, NM, USA, 2012.
25. Gedeon, D. *Sage User's Guide (v11 Edition)*; Gedeon Associates: Athens, OH, USA, 2016.

26. Fusco, A.; Ward, W.; Swift, G.W. Two-sensor power measurements in lossy ducts. *J. Acoust. Soc. Am.* **1992**, *91*, 2229–2235.
27. Biwa, T.; Tashiro, Y.; Nomura, H.; Ueda, Y.; Yazaki, T. Experimental verification of a two-sensor acoustic intensity measurement in lossy ducts. *J. Acoust. Soc. Am.* **2008**, *124*, 1584–1590.
28. Wilson, D.K.; McIntosh, J.D.; Lamber, R.F. Forchhemer-type nonlinearities for high-intensity propagation of pure tones in air-saturated porous media. *J. Acoust. Soc. Am.* **1988**, *84*, 350–359.
29. McIntosh, J.D.; Zuroski, M.T.; Lambert, R.F. Standing wave apparatus for measuring fundamental properties of acoustic materials in air. *J. Acoust. Soc. Am.* **1990**, *88*, 1929–1938.
30. McIntosh, J.D.; Zuroski, M.T. Nonlinear wave propagation through rigid porous materials. I: Nonlinear parameterization and numerical solutions. *J. Acoust. Soc. Am.* **1990**, *88*, 1939–1949.
31. Roh, H.; Raspet, R.; Bass, H.E. Parallel capillary-tube-based extension of thermoacoustic theory for random porous media. *J. Acoust. Soc. Am.* **2007**, *121*, 1413–1422.
32. Zwikker, C.; Kosten, C.W. *Sound Absorbing Materials*; Elsevier: Amsterdam, The Netherlands, 1949; Chapter 2.
33. Petculescu, A.; Wilen, L.A. Lumped-element technique for the measurement of complex density. *J. Acoust. Soc. Am.* **2001**, *110*, 1950–1957.

© 2017 by the authors. Licensee MDPI, Basel, Switzerland. This article is an open access article distributed under the terms and conditions of the Creative Commons Attribution (CC BY) license (http://creativecommons.org/licenses/by/4.0/).

Article

Modeling of Heat Transfer and Oscillating Flow in the Regenerator of a Pulse Tube Cryocooler Operating at 50 Hz

Xiufang Liu [1], Chen Chen [1], Qian Huang [1], Shubei Wang [2], Yu Hou [1] and Liang Chen [1,*]

1 State Key Laboratory of Multiphase Flow in Power Engineering, Xi'an Jiaotong University, Xi'an 710049, China; liuxiufang@mail.xjtu.edu.cn (X.L.); chenchenc@stu.xjtu.edu.cn (C.C.); huangqian1993@stu.xjtu.edu.cn (Q.H.); yuhou@mail.xjtu.edu.cn (Y.H.)

2 Xi'an Jiaotong University Su Zhou Academy, Suzhou 215123, China; wangwsb@xjtu.edu.cn

* Correspondence: liangchen@mail.xjtu.edu.cn; Tel.: +86-29-8266-4921

Academic Editor: Artur J. Jaworski

Received: 4 April 2017; Accepted: 22 May 2017; Published: 5 June 2017

Abstract: The regenerator of the pulse tube refrigerator (PTR) operates with oscillating pressure and mass flow, so a proper description of the heat transfer characteristics of the oscillating flow in the regenerator is crucial. In this paper, a one-dimensional model based on Lagrangian representation is developed to simulate the oscillating flow in the regenerator of the PTR. The continuity equation, momentum equation and energy equation are solved iteratively using the SIMPLER algorithm. The Darcy-Brinkman-Forchheimer model is used in the momentum equation, and a thermal non-equilibrium model is implemented in the energy equation. Lagrangian representation is employed to describe the thermodynamics of fluid parcels while the Eulerian representation (control volume method) is adopted for the energy equation of the solid matrix. The boundary conditions are set as the periodic flow of the sine function. The thermodynamic parameters of the gas parcels are obtained, which reveal the critical processes of the heat transfer in the regenerator under oscillating flow. The performance of the regenerator with different geometries is evaluated based on the numerical results. The present study provides insight for better understanding the physical process in the regenerator of the PTR, and the proposed model serves as a useful tool for the design and optimization of the cryogenic regenerator.

Keywords: regenerator; oscillating flow; pulse tube cryocooler; ineffectiveness

---

1. Introduction

The use of very low temperatures in advanced technology for applications in medical science, electronic technics, computer technology, aerospace exploration and other fields requires high performance cryogenic systems of low cost, simple design, small vibrations, low noise production, long life and high efficiency [1]. The pulse tube refrigerator (PTR) has been widely used because of the absence of moving parts at cryogenic temperature [2]. As the most important component in a PTR system, the regenerator has the advantages of a large heat transfer area and enhanced fluid mixing [3]. Its oscillating flow characteristics play a critical role in the performance of a PTR, which has attracted extensive attention in both theoretical and experimental studies. Radebaugh et al. reported a series of numerical studies on the performance and optimization of regenerators of a 4 K cryogenic system [4–7]. They used the commercial software Regen 3.2 in the simulation and design. They obtained the temperature curves along the regenerator which were later demonstrated to be consistent with the experimental results [7]. The comparative study of He-3 and He-4 systems showed that He-3 was more like an ideal gas at low temperature, which indicated that the substitution of He-3 for He-4 helped to improve the performance of regenerator [5]. Using Regen 3.3, the influence of porosity on the losses of the regenerator was analyzed, showing that the regenerator loss decreased with the decrease of porosity [4]. They also investigated the influence of the operating parameters, e.g., average pressure, pressure ratio and warm-end temperature; the regenerator parameters, e.g., material, shape and hydraulic diameter of the solid matrix; and the geometric parameters. Based on the optimization, the ineffectiveness was reduced to 0.36 and the second-law efficiency was enhanced to 25% [6]. Nield et al. [8] investigated the effects of a local thermal non-equilibrium on thermally-developing forced convection in a porous medium in parallel plate channels with the classic Graetz method, and the momentum equation of porous medium was established using the Brinkman model. Their study revealed that the local Nusselt number was closely related to the Peclet number and the thermal conductivity of both solids and gas. Pathak et al. [9] used the commercial computational fluid dynamics code (CFD) Fluent to simulate convective heat transfer and thermal dispersion during laminar pulsating flow in porous media, wherein the gas pulsation was realized through the inlet boundary condition of sine waves. Their simulations covered the porosities range from 0.64 to 0.84, the frequency range from 0 to 100 Hz, and the Reynolds number range from 70 to 980, which showed that the heat transfer in the porous medium had a strong dependence on the porosity, Reynolds number and operating frequency. Based on the simulation results, empirical correlations of both the thermal diffusion term and the Nusselt number were proposed, respectively. Teamah et al. [10] carried out numerical simulations of laminar forced convection in horizontal pipes partially or completely filled with porous material, and obtained the effects of the geometry (cylindrical and annular) of the porous media and the placement in the pipe on the flow distribution. The influence of the external diameter of porous media and the Darcy number was also investigated. The results showed, that the average Nusselt number of the flow in the pipes with the placement of porous media increased, while the entrance length decreased; there existed a critical value for the external diameter of porous media, and the thermal performance degraded when the external diameter exceeded this value. Li et al. [11] performed both theoretical and experimental studies to reveal the influence of regenerator void volume on the performance of a precooled 4 K high frequency Stirling type PTR. They found the following results: that the refrigeration temperature could be lowered as porosity increased, but that the power input of the precooling stage increased significantly; that lower average pressure yielded a lower refrigeration temperature, as well as a lower power input of the second precooling stage; that the performance of the final stage regenerator had a strong dependence on the operating frequency, especially at a low average pressure; and that the regenerator void volume in a multi-stage Stirling type PTR had a significant impact on the distribution of cooling capacity over different stages of regenerator. Saat and Jaworski [12] proposed a friction factor correlation applicable to a larger mesh range for a regenerator in a travelling-wave thermoacoustic system, based on experimental and CFD studies. They also studied the effects of temperature field, gravity and device orientation on the oscillatory flow [13]. Hsu and Biwa [14] carried out an experimental study on the heat flow in different mesh screen regenerators and compared the experimental result with four empirical equations, and concluded that all four empirical equations reproduce the measured heat flow rates to a similar degree.

In the previous studies, Eulerian representation has been widely employed to describe the oscillating flow and heat transfer in the regenerators. Recently, Dai and Yang [15,16] applied the lattice Boltzmann method (LBM) to the study of porous media flow and heat transfer. They verified the LBM model by simulating the flow across a cylinder, and then used the model to investigate the oscillating flow and heat transfer in porous media. The results showed that LBM was an effective method for the optimization of the PTR. In previous studies, the Lagrange description-based model has been adopted. In 1994, Organ [17] employed the Lagrangian representation for the governing equations of energy conservation. They solved the regenerator problem using the ε-NTU (number of transfer units) method, and analyzed the effect of NTU and solid-gas heat capacity ratio on the performance of the regenerator. However, the momentum equations were ignored in their study, and details of the oscillating flow in regenerators are still not well understood. Liang et al. [18] used the Lagrange description to explain the working mechanism of PTR and proposed the thermodynamic asymmetry effect. However, they focused on the pulse tube, and the regenerator was assumed to be perfect. Ataer [19] concluded that the Lagrangian method was more suitable for analyzing the heat transfer in the regenerative duct of free-piston type Stirling engines than other theories. In 2014, Tang et al. [20] developed a one-dimensional model of a regenerator based on the Lagrange description, and investigated the enthalpy flow, entropy flow on the Euler description and the gas parcels motion and loss on the Lagrange description, respectively. This work is helpful in understanding the regenerative process, but the flow resistance and real gas properties are neglected, and the heat transfer coefficient is also set constant.

In this study, Lagrangian representation is used to describe the thermodynamic cycle process of gas parcels inside the regenerator, and track the flow and heat transfer of a single parcel, leading to a better understanding of the operation principles and characteristics of the regenerator. A one-dimensional transient model of the regenerator in the PTR operating at 50 Hz is developed based on Lagrangian representation. The SIMPLER algorithm is used to solve the governing equations. The porous Darcy-Brinkman-Forchheimer model is applied in the momentum equation, and the thermal non-equilibrium model is implemented in the energy equation. The simulation results of mass flow at the cold end agree well with the experimental data, which verifies the reliability of this model. The variations of thermal parameters and the distribution of gas parcels under oscillating flow are obtained and analyzed in detail, and the effects of structural parameters on the regenerator ineffectiveness are discussed. This study is useful in aiding the understanding of the flow and heat transfer of internal fluid inside the regenerator, and can also provide the guidelines for the regenerator design.

2. Numerical Model and Simulations

2.1. Mesh Generation

A one-dimensional transient model of the regenerator under oscillating flow is developed. Considering the use of Lagrangian representation, it is necessary to describe the mesh generation before discussing the governing equations. Using Lagrangian representation, the fluid in the regenerator is divided into parcels, and the governing equations of gas are set up to track the motion, deformation and heat transfer processes of gas parcels in the regenerator. A "staggered grid" scheme is implemented for the pressure and velocity, as shown in Figure 1 for gas parcel i. It should be noted that the staggered grid of the parcels is different from that of control volumes. In the Lagrangian representation, the velocity and the deformation are defined at the parcel boundary, while the pressure, temperature and other properties are defined at the center of the parcels. The parcel boundaries move forward and backward in the simulation. However, the staggered grid is fixed in the control volume scheme. The flow direction from the warm end to the cold end in the regenerator is defined as positive, and the orders of parcels along the positive direction at the boundary are defined as their indices. The Eulerian representation is used for the mesh generation of the solid matrix in the regenerator, and the energy equations for control volumes are solved using the finite volume method.

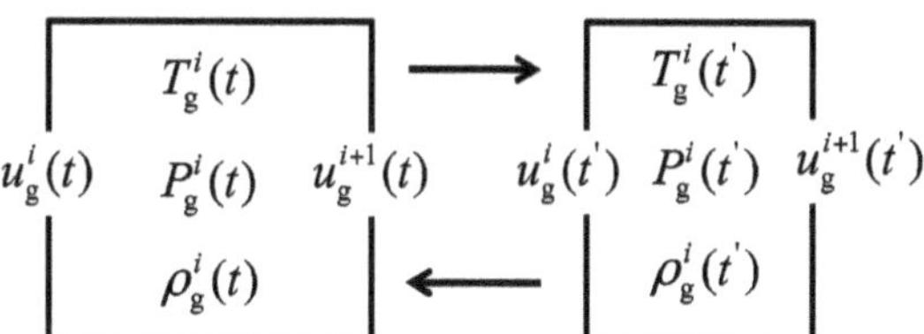

Figure 1. Velocity and thermodynamic parameters of gas parcels in the regenerator model: T_g—emperature of gas parcel; P_g—pressure of gas parcel; ρ_g—density of gas parcel; u_g—velocity of gas parcel; t—instant time; i—number of gas parcel.

2.2. Governing Equations

2.2.1. Assumptions

1. Heat transfer and fluid flow is simplified to one dimension along the axial direction of the regenerator;
2. The wall of the regenerator is adiabatic;
3. Axial conduction in the solid matrix is neglected;
4. Heat conduction between the wall and solid matrix in the regenerator is negligible;
5. Density of the solid matrix is assumed to be constant and thermal expansion of the solid is neglected.

2.2.2. Mass, Momentum and Energy Conservation

The governing equations of the conservations of mass, momentum and energy are presented in Lagrangian form as follows:

The continuity equation of gas parcels is:

$$\frac{D\rho}{Dt} + \rho\frac{\partial u}{\partial x} = 0 \tag{1}$$

The momentum equation is based on the Darcy-Brinkman-Forchheimer model:

$$\rho\frac{Du}{Dt} = -\frac{\partial p}{\partial x} + S \tag{2}$$

where S is the source term representing the force between fluid and wire meshes. It includes the osmotic force term and inertial force term by:

$$S = -(C_f\mu + \frac{1}{2}C_j\rho u)u \tag{3}$$

where C_f is the osmotic resistance coefficient and C_j is the inertia coefficient. The two coefficients can be estimated according to the friction factor relations of the regenerator, which will be discussed in the following section.

Energy equations of the gas parcels and solid matrix are respectively written based on the thermal non-equilibrium model. For the control volume, the heat transferred to gas by heat convection is equal to the sum of the enthalpy increment of flow and the internal energy increment. For the gas parcels in Lagrangian representation, the heat transferred to gas by heat convection equates to the increment of internal energy of the gas parcel itself. So the expression of energy conservation is given by:

$$h\{T_w(x,t) - T_g(x,t)\}Pdx = \rho c_v A_x dx\left\{\frac{\partial T_g(x,t)}{\partial t}\right\} \tag{4}$$

For the solid matrix, the heat from the gas equates to the internal energy increment of the solid matrix, and the expression of energy conservation equation is as follows:

$$h\{T_g(x,t) - T_w(x,t)\}Pdx = \rho_w c_w A_x dx\frac{1-K}{K}\frac{\partial T_w(x,t)}{\partial t} \tag{5}$$

2.3. Boundary Conditions

In order to simulate the oscillating flow in the regenerator, a sinusoidal pressure wave is provided at the cold end with an average pressure of 3 MPa, and the mass flow of sinusoidal variations is provided at the warm end with the magnitude of 2.5 g/s. The thermal boundary conditions are as follows: the inlet temperature at the cold end is 80 K; the inlet temperature at the warm end is 285 K. Thus, the boundary conditions are specified as:

$$\begin{aligned} m &= m_{\mathrm{h}},\ T = T_{\mathrm{h}},\ at\ x = 0 \\ P &= P_{\mathrm{c}},\ T = T_{\mathrm{c}},\ at\ x = \mathrm{L} \end{aligned} \tag{6}$$

where subscript h indicates the warm end and c is the cold end.

2.4. Correlations of Heat Transfer Coefficient and Friction Factor

In 2014, Coasta et al. [21] reported the correlation of heat transfer coefficient for the stacked woven wire matrix regenerator, which had a diameter of 100 μm and a porosity of 63%. The validity and reliability of this correlation has been demonstrated for the application of porous media simulations, and the expression is given by:

$$Nu = 1.91 + 0.17Re^{0.80} \tag{7}$$

The friction factor correlation has been widely studied for porous media. Miyabe et al. [22] investigated the pressure drop under steady flow in a screen-based regenerator and obtained the Darcy friction factor correlation, which is now widely used under steady flow in the regenerator:

$$f = \frac{33.6}{Re_l} + 0.337 \tag{8}$$

where l denotes the pore gap.

Nam et al. [23] investigated the oscillatory flow in the regenerator at ambient temperature and set up a one-dimensional model. Based on the experimental results, they proposed correlations for the Fanning friction factor, which depended on the mesh numbers:

$$\begin{aligned} f_{\mathrm{oscm}} &= 36.55/Re_{\mathrm{m}} + 0.16 \quad \text{for mesh 200} \\ f_{\mathrm{oscm}} &= 44.10/Re_{\mathrm{m}} + 0.33 \quad \text{for mesh 250} \\ f_{\mathrm{oscm}} &= 48.83/Re_{\mathrm{m}} + 0.16 \quad \text{for mesh 400} \end{aligned} \tag{9}$$

In this study, the steady friction factor correlation by Miyabe et al and the oscillating flow friction factor correlation by Nam et al. are applied in the simulations, respectively. The results indicate that under the boundary condition of sinusoidal pressure, the steady friction factor correlation by Miyabe is no longer valid, showing significant differences between steady flow and oscillatory flow. So, the friction factor correlation introduced by Nam et al. is selected.

2.5. Numerical Algorithm

The numerical algorithm to solve the governing equations is shown in Figure 2. The geometry parameters are given at the beginning. The boundary conditions and initial data are set to start the calculation. An iteration at a given time step involves calculating the velocity, pressure and temperature of all gas parcels as well as the temperature of all solid elements. After each iteration, the gas velocity of each parcel is compared with the value in the previous iteration, thus obtaining the maximum absolute difference in all gas parcels. This value is then compared with the convergence criteria. If the convergence condition is not satisfied, iterations in the same time step are continued with the updated velocity, pressure and temperature. When the convergence criteria are achieved, the iteration in this time step is completed and the boundary conditions are updated. The calculation in the next time step starts with the updated values of gas velocities, pressures, temperatures, and the new temperature field of solid matrix.

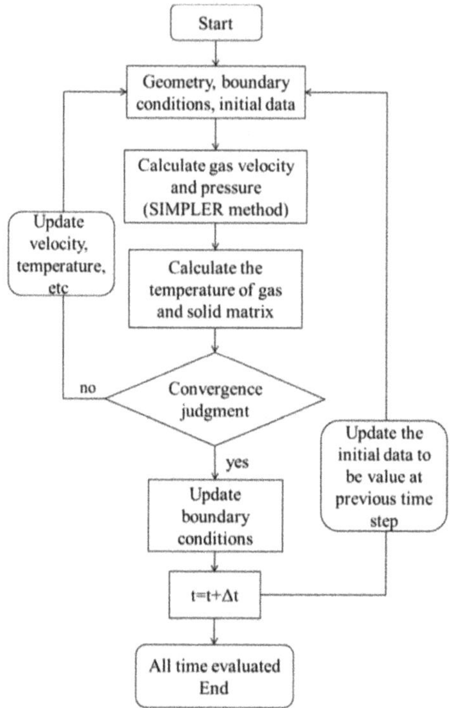

Figure 2. Schematic of the numerical algorithm.

3. Results and Discussion

3.1. Mesh Independence Verification

The mesh independence is verified for the selection of time step and grid size, and the regenerator parameters used in the simulations are listed in Table 1. Firstly, the selection of the time step is verified. Simulations with different time steps are performed under the condition of the fixed parcel grids, and the effects of the time step on the results are compared. The proper time step is the one beyond which further refinement has little effect on the results. Then the selection of the size of gas parcels is verified through simulations with a fixed time step, but different initial sizes of gas parcels. The proper grid is obtained when further mesh refinement does not improve the simulation results.

Table 1. Typical geometry parameters of the regenerator in the mesh independent verification.

Geometry	Parameters
length	65 mm
inner diameter	17.5 mm
porosity	0.7
mesh number	400
hydraulic diameter of mesh wire	43 μm

3.1.1. Verification of Time Step Independence

Four different time steps under the fixed initial size of gas parcels are chosen as follows: 50, 100, 200 and 300 time steps per cycle. The variations of inlet velocities at the cold end are obtained and shown in Figure 3, serving as the selection criteria of time steps.

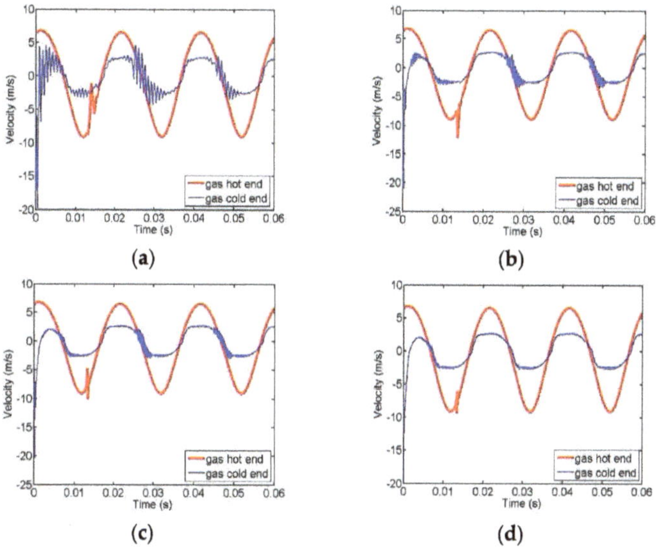

Figure 3. Velocity variations at the cold end of the regenerator predicted using different time steps per cycle: **(a)** 50; **(b)** 100; **(c)** 200; **(d)** 300. The velocity at the warm end is also shown in each figure as reference.

As shown in Figure 3, the red solid lines represent the specified velocity at the warm end, and the blue lines are the predictions of velocities at the cold end. From these 4 figures, it can be seen that the calculated value of cold end velocity is unstable, with large fluctuations when a cycle is divided into 50 time steps; the fluctuation of the calculated cold end velocity becomes smaller under the condition of 100 time steps per cycle; a division of 200 time steps per cycle shows even better stability and lower fluctuations; the choice of 300 time steps per cycle enhances the stability of the calculated velocity at the cold end and minimizes the fluctuation to a great extent. As shown in Figure 4, the cold end velocity after several cycles converges with little fluctuations under the condition of 200 time steps per cycle, which is almost the same as under 300 time steps per cycle. At the early stage, the initial velocity, which is inconsistent with actual condition, produces a large fluctuation. Therefore, a division of 200 time steps per cycle is adopted in the consideration of both the calculation accuracy and computational cost.

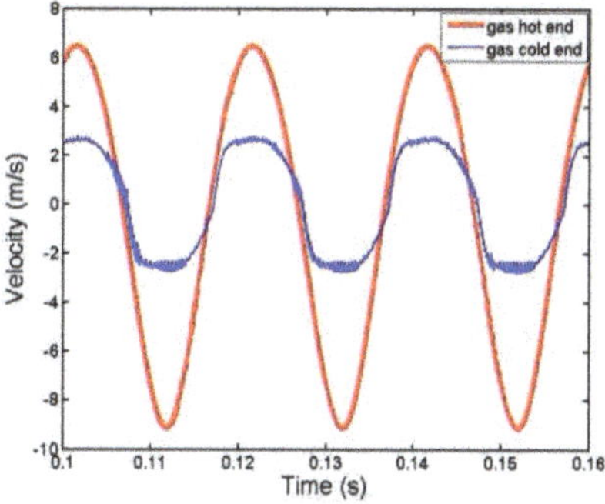

Figure 4. Predictions of cold end velocity after several cycles when time steps = 200 per cycle.

3.1.2. Size of Parcels Independent Verification

Under the condition of fixed time steps, the initial gas in the regenerator is divided into 65, 130, 195 parcels, and the corresponding parcel sizes are 1, 0.5 and 0.333 mm, respectively. Then, simulations for a few cycles are performed, and the predictions of cold end velocities are compared, which enables us to make a proper choice of parcel meshing.

Figure 5 illustrates that, when the initial gas parcels inside the regenerator increase from 65 to 130, the predictions of the cold end velocity are converged when the initial parcel number is 130; the predictions show little change with further increase of the initial parcel number beyond 130. Therefore, the parcel meshing of 130 is chosen in the following simulations.

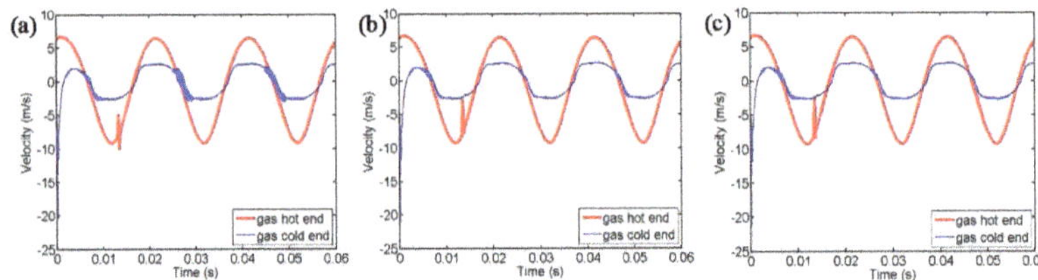

Figure 5. Predictions of cold end velocity for different initial parcel sizes: (**a**) 1 mm; (**b**) 0.5 mm; (**c**) 0.333 mm. A cycle is divided into 200 time steps for all the simulations.

3.2. Model Validation

Nam et al. [24] conducted an experimental study on regenerators within a given size and operating parameters. The comparison between the results calculated by our model and that of the experiment is carried out with the structural and operating parameters of the regenerator in the experiment. The proposed model is also compared with Regen, a program specialized in regenerator simulation [25]. The pressure variations of the warm end and the cold end in one cycle are compared as shown in Figure 6. In general, the simulation results are compatible with the experimental results, more so than Regen. The deviation of Regen may be caused by the limitations in the setting of boundary conditions, namely that both pressure and mass flow rate must be specified at the cold end, but that the input value of the phase angle between the two may be deviated from the actual value. Another reason might be because of the friction term determined from the experimental data of steady flow, which may not be accurate for oscillating flow in the regenerator. It can also be seen from the figure that the deviations in amplitudes and phase angles between the pressure curves predicted by our model, and the experimental results of both the warm end and the cold end, are relatively small. This deviation may be caused by the assumptions in the model, as well as the difference in boundary conditions at the cold end as denoted by the blue curves in Figure 6.

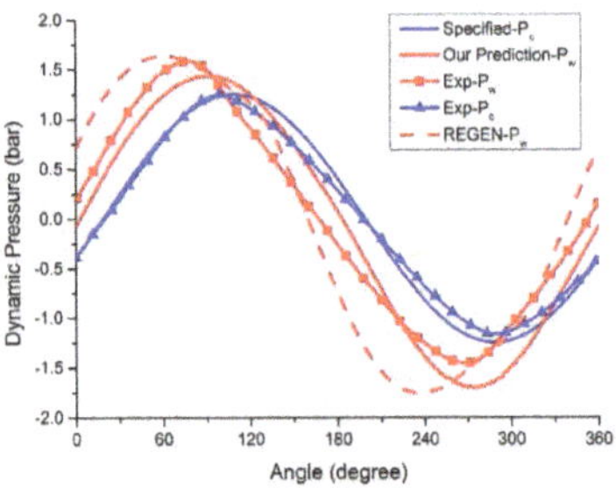

Figure 6. Comparison of the predicted pressure variations with the experimental results [24] and the calculated results by Regen [25].

3.3. Simulation Results

A high frequency regenerator at low temperature, operating at 50 Hz with a cold end temperature of 80 K, and a warm end temperature of 285 K, is simulated. The flow field and heat transfer in the oscillating processes are discussed in detail. With the aid of Lagrangian representation applied in the model, the trajectories of selected gas parcels inside the regenerator can easily be obtained, as well as such time-dependent changes of gas parcel parameters as temperature, pressure, density and velocity.

The trajectories of gas parcels inside the regenerator are illustrated in Figure 7. The gas parcels are numbered on the basis of their positions at the initial time, with indices increasing sequentially along the direction from warm- to cold-ends. The five parcels numbered 25, 50, 75, 100 and 125 are selected for tracking purposes. It can be seen from the trajectories of gas parcels that the parcels' backward and forward motion over time shows an oscillatory pattern of movement, and that the parcels close to the warm end would be expelled from the regenerator at the second half cycle. From the figure, it can be observed that the highest and lowest points during the time period of the gas parcels' trajectories decrease gradually, which indicates the gas parcels approaching the warm end in the first few cycles of the simulation.

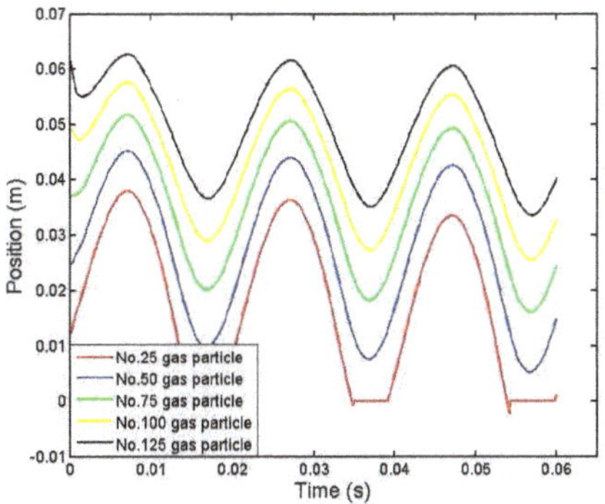

Figure 7. The trajectories of selected gas parcels in the regenerator.

The variations of temperature, pressure, density and velocity with the positions of the five gas parcels tracked in the regenerator are shown in Figure 8. It can be established from the figures that each gas parcel's trajectory consists of three parts of curves, which is consistent with the three cycles of the simulation, indicating that gas parcels move in a reciprocating pattern. The temperature variations shown in Figure 8a indicate that the temperature decreases when gas parcels move from the warm end to the cold end and the temperature rises when moving from the cold end to the warm end. The average temperature of the first half of the cycle is higher than that of the second half, which clearly illustrates the regenerator's cooling mechanism; the parcel's average temperature decreases with the increase of its index that is ordered from the warm end to the cold end. These gas parcels form a series of heat exchange units at different temperatures from 285 to 80 K. These results help to improve our understanding of how regenerators work from the perspective of gas parcels.

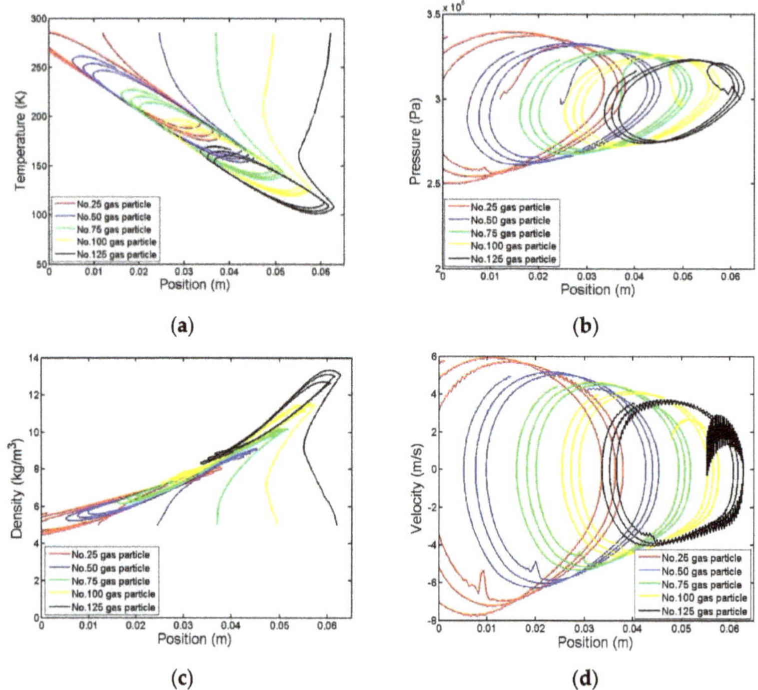

Figure 8. Parameter variations of selected gas parcels as a function of position: (**a**) Temperature; (**b**) Pressure; (**c**) Density; (**d**) Velocity.

Figure 8b shows the pressure variations of selected gas parcels as a function of position. The parcels' pressure firstly increases and then decreases at a rather slow speed when moving from the warm end to the cold end, while the opposite is true when the process is in reverse, moving from the cold end to the warm end. The average pressure in the first half cycle is higher than that in the second half. Under the influence of temperature and pressure, the parcels' density shown in Figure 8c rises constantly during the motion from the warm end to the cold end and decreases when moving in the opposite direction. The average density of the cooling process is higher than the heating process. The variations of parcel velocity, as shown in Figure 8d, follow a trend similar to the pressure. When combining the results from the four figures, it is obvious that the variations of the parcels' temperature, pressure, density and velocity in different cycles decrease with the increase of parcel index: in other words, gas parcels near the warm end change most dramatically in the first three cycles. This may result from the fact that the parcel's velocity is larger when mass flow is constant with a rather smaller gas density under the higher temperature at the warm end. The larger velocity can result in a larger range of motion within the same time cycle, thus causing larger resistance and pressure differences, and which therefore leads to dramatic velocity variations. Meanwhile, gas volume is considerably large when the gas density is small, since the mass of each gas parcel at the initial time is the same, thus increasing the heat exchange area between parcels and solid matrix and enlarging the temperature differences. The large variations of parcels' temperature and pressure yield a large variation of parcel density.

According to the Reynolds transport theorem, Lagrangian representation can be converted to Eulerian representation. Figure 9 shows the distribution of parameters inside the regenerator from the perspective of a Eulerian representation. It can be seen from the distribution of gas temperature that the temperature declines from the warm end to the cold end. With the initial value of 285 K, the heat exchange increases, and the gas temperature decreases gradually in the process of calculation. It is noticeable from Figure 9a that the temperature drops dramatically near the cold end during the simulation of the cooling-down processes. This is because the temperature of the gas entering the regenerator at the cold end is set to 80 K during the second half cycle of the oscillating flow. In the early stages, the large temperature difference between gas parcels and solid matrix results in dramatic temperature change. As the oscillating flow continues, the distribution of temperature develops into a linear pattern. As shown in Figure 9b, the distribution of pressure in the regenerator changes periodically over the cycle. During the first half cycle of the oscillating gas flow, pressure declines from the warm end to the cold end, while during the second half cycle, the pressure distribution is reversed. The distribution of density shown in Figure 9c is similar to the temperature distribution on the existence of dramatic changes, which can result from the temperature variations. As shown in Figure 9d, the velocity seems nearly linear, with apparent fluctuations near the cold end. This phenomenon can be explained by the reversals of gas parcels. There are gas parcels with different speeds and directions flowing into both ends of the regenerator during the entire cycle of the oscillating flow, which generates an unstable pressure in the process of parcel velocity reversal, and therefore leads to the presentence of velocity fluctuations. Since the parcels from the warm end have higher average velocity, the velocity reversal occurs near the cold end, which can be observed in Figure 9d.

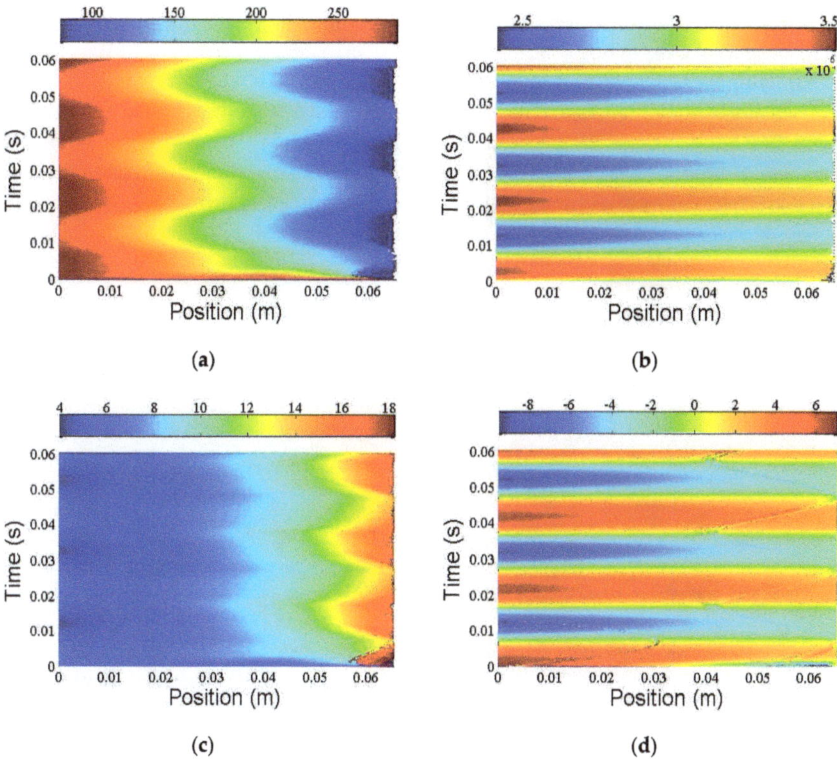

Figure 9. Distribution of gas parameters in the regenerator: (**a**) Temperature; (**b**) Pressure; (**c**) Density; (**d**) Velocity.

The thermal performance of the regenerator is usually measured by the ineffectiveness, which is as follows [26]:

$$\lambda \equiv \frac{\left(\int_0^{\tau} mhdt\right)_C}{\left(\int_0^{\tau_w} mhdt\right)_W + \left(\int_{\tau_w}^{\tau} mhdt\right)_C} = \frac{H}{H_W - H_C} \tag{10}$$

where $0 - \tau_w$ is the time period when fluid flows into the warm end, and $\tau_w - \tau$ is the time period when fluid flows into the cold end. H represents the enthalpy loss at the cold end during an entire cycle, and $H_w - H_c$ represents the maximum heat transfer that can be achieved in the regenerator.

The purpose of the regenerator is to deliver as much of this acoustic power to the cold end as possible with minimum loss. Therefore, a net refrigeration power is given by:

$$\dot{Q}_{net} = \left\langle P\dot{V} \right\rangle_c - \left\langle \dot{H} \right\rangle_p - \dot{Q}_{reg} - \dot{Q}_{cond} - \dot{Q}_{pt} - \dot{Q}_{rad} \tag{11}$$

where $\left\langle P\dot{V} \right\rangle_c$ is the acoustic power at the cold end; $\left\langle \dot{H} \right\rangle_p$ is the enthalpy flow associated with the enthalpy pressure dependence (real gas effect); $\dot{Q}_{reg}$ is the thermal loss associated with enthalpy flow caused by imperfect heat transfer and limited heat capacity in the regenerator (regenerator ineffectiveness); $\dot{Q}_{cond}$ is the conduction heat leak; $\dot{Q}_{pt}$ is the loss associated with an imperfect pulse tube or any irreversible expansion process at the cold end; and $\dot{Q}_{rad}$ is the radiation heat leak to the cold end, which is ignored in this work.

Considering the lack of accurate theoretical calculation for pulse tube losses, an estimated value of pulse tube losses is adopted, as the pulse tube expansion efficiency is assumed to be 80%. The loss characteristics under our operating conditions are calculated and the proportions of various losses in $\left\langle P\dot{V} \right\rangle_c$ are compared. Figure 10 gives the distribution of losses in a PTR, which indicates that regenerator ineffectiveness is one of the main reasons contributing towards refrigeration losses.

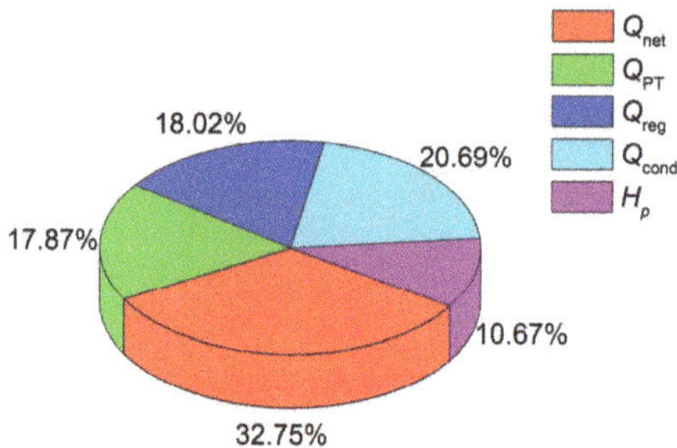

Figure 10. Contribution to the system losses from real gas effect, regenerator ineffectiveness, conduction heat leak and irreversible expansion.

By using the above correlation, the ineffectiveness of regenerators with different diameters and lengths are estimated to analyze their effects on the regenerator performance. With respect to the effect of diameter on the regenerator performance, five regenerators are studied, with diameters of 12, 14, 16, 17.5 and 20 mm, respectively, and the results are shown in Figure 11. With the increase of diameter, the ineffectiveness of the regenerator shows a tendency towards increase at a smaller diameter, reaching the maximum near the diameter of 14 mm, and then decreases quickly before being stabilized at the diameter of 20 mm. The occurrence of the ineffectiveness peak can be explained as follows: the heat exchange between fluid and solid matrix is associated with heat transfer coefficient and heat transfer area. On the basis of this principle, under the condition of constant mass flow, a large fluid velocity accompanied by a small diameter, leads to a large heat transfer coefficient and thus minimal ineffectiveness; a large heat transfer area accompanied by a large diameter can also reduce the ineffectiveness; when the diameter of regenerator is intermediate, the combination of fluid velocity and heat transfer area would lead to the worst heat exchange, and as a result, the ineffectiveness reaches the maximum value.

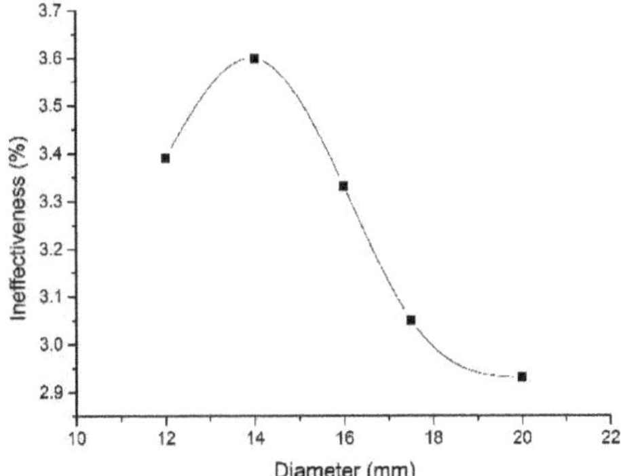

Figure 11. Ineffectiveness variations as a function of regenerator diameter.

In order to obtain the effect of the length on the regenerator performance, simulations are performed on five regenerators with different lengths: 20, 40, 65, 80 and 100 mm, and the results are shown in Figure 12. It can be observed that the ineffectiveness decreases monotonically with increasing length of the regenerator. The variations become smaller with further increase of length beyond 65 mm, and finally tend to stabilize at the length of 100 mm. It can be explained as follows: the larger heat transfer area of the longer-length regenerator enhances the heat transfer, and thus, reduces the ineffectiveness. When the regenerator length is large enough, the heat transfer between fluid and solid matrix has reached the limit, and any further increase of length contributes little to the heat transfer enhancement, as well as the ineffectiveness reduction.

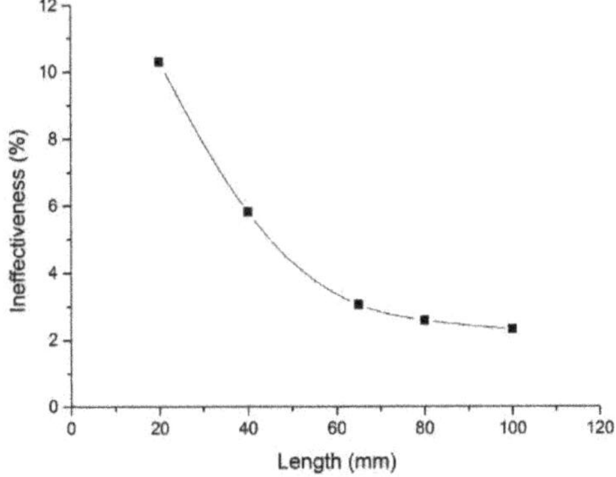

Figure 12. Ineffectiveness variations as a function of regenerator length.

4. Conclusions

Simulations of oscillating flow in a pulse tube cryocooler regenerator operating at 50 Hz were performed, using a numerical model in Lagrangian representation, and solved with SIMPLER algorithm. The study reveals that the flow resistance under oscillating flow differs significantly from that which occurs under steady flow, indicating that the friction factor correlation for steady flow was not valid under the condition of oscillating flow. The trajectory obtained by tracking a single gas parcel implies that the oscillating movement in the regenerator can provide insight into understanding the complex flow. The analysis on the temperature variations of gas parcels reveals the heat exchange processes in the regenerator. Additionally, similar characteristics exist in the variations of gas parcels' temperature, pressure, density and velocity, that is, the variation range of parameters near the warm end is larger, due to the higher temperature and smaller density. According to the Reynolds transport theorem, the distributions of various parameters along the axial direction of the regenerator are obtained from the perspective of the Euler representation. Further study of flow and heat transfer processes shows that the trends of temperature, pressure and density under the initial conditions are compatible with the general characteristics of the regenerator. However, the distribution of velocity shows some fluctuations near the cold end, which is caused by the inlet velocity difference at the warm- and cold-ends. The performance of the regenerator with different diameters and lengths is evaluated, showing that the ineffectiveness reaches its peak when the diameter is near 14 mm and then decreases with the increase of diameter; the ineffectiveness decreases monotonically with increasing length of the regenerator, and tends to stabilize at the length of 100 mm.

Acknowledgments: This project was supported by the National key basic research program (613322), and the Natural Science Foundation of Jiangsu Province (BK20160388).

Author Contributions: Liang Chen and Chen Chen conceived and developed the model; Xiufang Liu and Chen Chen performed the simulations; Chen Chen and Yu Hou analyzed the data; Shubei Wang contributed analysis tools; Xiufang Liu and Qian Huang wrote the paper.

Conflicts of Interest: The authors declare no conflict of interest. The founding sponsors had no role in the design of the study; in the collection, analyses, or interpretation of data; in the writing of the manuscript, and in the decision to publish the results.

References

1. Popescu, G.; Radcenco, V.; Gargalian, E.; Bala, P.R. A critical review of pulse tube cryogenerator research. *Int. J. Refrig.* **2001**, *24*, 230–237. [CrossRef]
2. Richardson, R.N.; Evans, B.E. A review of pulse tube refrigeration. *Int. J. Refrig.* **1997**, *20*, 367–373. [CrossRef]
3. Mahdi, R.A.; Mohammed, H.A.; Munisamy, K.M.; Saeid, N.H. Review of convection heat transfer and fluid flow in porous media with nanofluid. *Renew. Sustain. Energy Rev.* **2015**, *41*, 715–734. [CrossRef]
4. Radebaugh, R.; Huang, Y.; O'Gallagher, A. Calculated Performance of Low-Porosity Regenerators at 4 K with He4 and He3. *Cryocooler* **2009**, *15*, 325–334.
5. Radebaugh, R.; Huang, Y.; O'Gallagher, A.; Gary, J. Calculated regenerator performance at 4 k with helium-4 and helium-3. In Proceedings of the AIP Conference, Chattanooga, TN, USA, 16–20 July 2007.
6. Radebaugh, R.; Huang, Y.; O'Gallagher, A.; Gary, J. Optimization Calculations for a 30 HZ, 4 K Regenerator with HELIUM-3 Working Fluid. In Proceedings of the AIP Conference, Tucson, AZ, USA, 28 June–2 July 2009.
7. Radebaugh, R.; O'Gallagher, A.; Gary, J. Regenerator behavior at 4 K: Effect of volume and porosity. In Proceedings of the AIP Conference, Madison, WI, USA, 16–20 July 2001.
8. Nield, D.A.; Kuznetsov, A.V.; Xiong, M. Effect of local thermal non-equilibrium on thermally developing forced convection in a porous medium. *Int. J. Heat Mass Transf.* **2002**, *45*, 4949–4955. [CrossRef]
9. Pathak, M.G.; Ghiaasiaan, S.M. Convective heat transfer and thermal dispersion during laminar pulsating flow in porous media. *Int. J. Therm. Sci.* **2011**, *50*, 440–448. [CrossRef]
10. Teamah, M.A.; El-Maghlany, W.M.; Dawood, M.M.K. Numerical simulation of laminar forced convection in horizontal pipe partially or completely filled with porous material. *Int. J. Therm. Sci.* **2011**, *50*, 1512–1522. [CrossRef]
11. Li, Z.; Jiang, Y.; Gan, Z.; Qiu, L.; Chen, J. Influence of regenerator void volume on performance of a precooled 4 K stirling type pulse tube cryocooler. *Cryogenics* **2015**, *70*, 34–40. [CrossRef]
12. Saat, F.M.; Jaworski, A.J. Friction Factor Correlation for Regenerator Working in a Travelling-Wave Thermoacoustic System. *Appl. Sci.* **2017**, *7*, 253. [CrossRef]
13. Saat, F.M.; Jaworski, A.J. The Effect of Temperature Field on Low Amplitude Oscillatory Flow within a Parallel-Plate Heat Exchanger in a Standing Wave Thermoacoustic System. *Appl. Sci.* **2017**, *7*, 417. [CrossRef]

14. Hsu, S.; Biwa, T. Measurement of Heat Flow Transmitted through a Stacked-Screen Regenerator of Thermoacoustic Engine. *Appl. Sci.* **2017**, *7*, 303. [CrossRef]
15. Dai, Q.; Yang, L. LBM numerical study on oscillating flow and heat transfer in porous media. *Appl. Therm. Eng.* **2013**, *54*, 16–25. [CrossRef]
16. Dai, Q.; Chen, H.; Yang, L. Numerical simulations of oscillating flow and heat transfer in porous media by lattice boltzmann method. *J. Eng. Thermophys.* **2012**, *32*, 1891–1898.
17. Organ, A.J. Solution of the Classic Thermal Regenerator Problem. *ARCHIVE Proc. Inst. Mech. Eng. Part C J. Mech. Eng. Sci.* **1994**, *208*, 187–197. [CrossRef]
18. Liang, J.; Ravex, A.; Rolland, P. Study on pulse tube refrigeration Part 1: Thermodynamic nonsymmetry effect. *Cryogenics* **1996**, *36*, 87–93. [CrossRef]
19. Ataer, Ö.E. Numerical analysis of regenerators of free-piston type Stirling engines using Lagrangian formulation. *Int. J. Refrig.* **2002**, *25*, 640–652.
20. Tang, Q.; Cai, J.; Liu, Y.; Chen, H. A Numerical Model of Regenerator Based on Lagrange Description. In Proceedings of the Interational Cryocooler Conference, Syracuse, NY, USA, 9–12 June 2014.
21. Costa, S.C.; Barreno, I.; Tutar, M.; Esnaola, J.A.; Barrutia, H. The thermal non-equilibrium porous media modelling for CFD study of woven wire matrix of a Stirling regenerator. *Energy Convers. Manag.* **2015**, *89*, 473–483. [CrossRef]
22. Miyabe, H.; Hamaguchi, K.; Takahashi, K. An approach to the design of Stirling engine regenerator matrix using packs of wire gauzes. In Proceedings of the 17th Intersociety Energy Conversion Engineering conference, Los Angeles, CA, USA, 8 August 1982.
23. Nam, K.; Jeong, S. Novel flow analysis of regenerator under oscillating flow with pulsating pressure. *Cryogenics* **2005**, *45*, 368–379. [CrossRef]
24. Nam, K.; Jeong, S. Measurement of cryogenic regenerator characteristics under oscillating flow and pulsating pressure. *Cryogenics* **2003**, *43*, 575–581. [CrossRef]
25. Gary, J.; Radebaugh, R. An improved numerical model for calculation of regenerator performance (regen3.1). In Proceedings of the Fourth Interagency Meeting on Cryocoolers, Plymouth, MA, USA, 24 October 1990.
26. Radebaugh, R.; Linenberger, D.; Voth, R.O. Methods for the measurement of regenerator ineffectiveness. *NBS Spec. Publ.* **1981**, *607*, 70–81.

© 2017 by the authors. Licensee MDPI, Basel, Switzerland. This article is an open access article distributed under the terms and conditions of the Creative Commons Attribution (CC BY) license (http://creativecommons.org/licenses/by/4.0/).

Article

Friction Factor Correlation for Regenerator Working in a Travelling-Wave Thermoacoustic System

Fatimah A. Z. Mohd Saat [1] and Artur J. Jaworski [2,*]

[1] Centre for Advanced Research on Energy, Faculty of Mechanical Engineering, Universiti Teknikal Malaysia Melaka, Hang Tuah Jaya, 76100 Durian Tunggal, Melaka, Malaysia; fatimah@utem.edu.my

[2] Faculty of Engineering, University of Leeds, Woodhouse Lane, Leeds LS2 9JT, UK

* Correspondence: a.j.jaworski@leeds.ac.uk; Tel.: +44-113-343-4871

Academic Editor: Vitalyi Gusev

Received: 25 January 2017; Accepted: 2 March 2017; Published: 5 March 2017

Abstract: Regenerator is a porous solid structure which is important in the travelling-wave thermoacoustic system. It provides the necessary contact surface and thermal capacity for the working gas to undergo a thermodynamic cycle under acoustic oscillatory flow conditions. However, it also creates a pressure drop that could degrade the overall system performance. Ideally, in a travelling-wave system, the phase angle between oscillating pressure and velocity in the regenerator should be zero, or as close to zero as possible. In this study, the hydrodynamic condition of a regenerator has been investigated both experimentally (in a purpose-built rig providing a travelling-wave phasing) and numerically. A two-dimensional ANSYS FLUENT CFD model, capturing the important features of the experimental conditions, has been developed. The findings suggest that a steady-state correlation, commonly used in designing thermoacoustic systems, is applicable provided that the travelling-wave phase angle is maintained. However, for coarse mesh regenerators, the results show interesting "phase shifting" phenomena, which may limit the correlation validity. Current experimental and CFD studies are important for predicting the viscous losses in future models of thermoacoustic systems.

Keywords: thermoacoustic system; acoustic travelling-wave; porous medium; regenerator hydrodynamic condition; phase angle; flow resistance losses

---

1. Introduction

The underlying mechanism of the thermoacoustic energy conversion processes is a heat transfer interaction taking place between the gas parcels undergoing oscillatory motion and a solid material along which the gas oscillations occur. In thermoacoustic prime movers (engines), a spontaneous acoustic wave is excited which causes the gas parcel oscillations coupled with their compression and expansion, thus providing a means of transporting heat from the hotter to the cooler place of the solid. In thermoacoustic coolers, the acoustically induced displacement/compression/expansion of gas parcels leads to heat pumping effects in the solid. The solid material used in such applications often takes the form of a porous structure—for example, stacked layers of woven mesh screens—which by analogy to classic Stirling engines is referred to as "regenerator".

Figure 1a explains the interaction of gas parcels with a solid material (in this example, a plate with an imposed temperature gradient). The gas experiences thermal expansion during the displacement towards the higher temperature and thermal contraction during the displacement towards the lower temperature. In this way, the correct time phasing is obtained to meet the Rayleigh's criterion [1] to produce power. Figure 1b illustrates the thermodynamic cycle in comparison to the classic ideal

Stirling cycle, while Figure 1c shows one of the known configurations of thermoacoustic engines referred to as Thermo-Acoustic Stirling Heat Engine, TASHE [2].

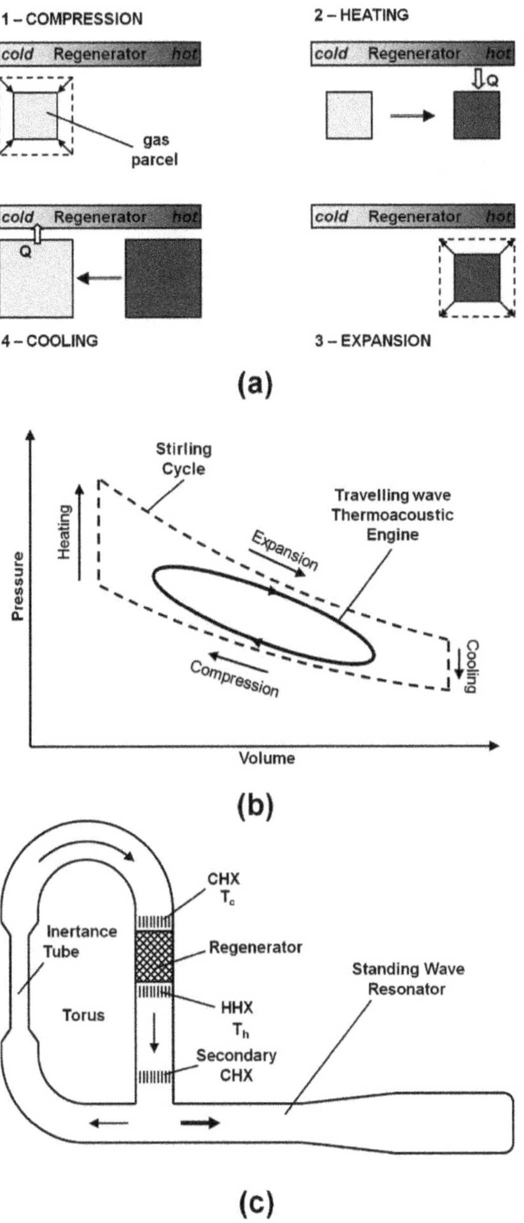

Figure 1. (**a**) Schematic of a gas parcel oscillating in the vicinity of a solid plate with an imposed temperature gradient, undergoing stages of compression and expansion; (**b**) pressure-volume (P-V) diagram illustrating the nature of the thermodynamic cycle; and (c) schematic of a possible configuration of a thermoacoustic travelling-wave engine.

Strictly speaking, the thermoacoustic core (a regenerator placed between the hot and cold heat exchangers) works simply as a power amplifier. To complete this Stirling-like thermodynamic process, acoustic power has to be fed to the ambient end of the regenerator with a near travelling-wave phasing. This leads to practical designs having feedback resonator types (closed loops) with various acoustic network elements such as inertance and compliance to control the correct/preferred phasing between pressure and velocity oscillation; further discussions are available in reference [3].

Thermoacoustic systems have gained considerable attention due to their environmental friendliness arising from the use of inert/noble gases as the working media and the lack of moving parts when executing the thermodynamic cycle which makes them of simple design and potentially of high reliability and low cost.

As part of the requirement to achieve high performance, the regenerator is used in travelling-wave thermoacoustic device with an aim to provide sufficient thermal contact with the working medium but also cause as little viscous loss as possible. Thermal contact is improved due to the large surface area of the porous and tortuous structure of the regenerator. However, care is needed to avoid high viscous losses. From the practical point of view, and to illustrate these problems, the regenerators are typically constructed from metal wire mesh screens by compacting a large number of wire mesh disks (tens or hundreds) along the direction of wave propagation. The conflict between efficient thermal contact and unwanted pressure losses of the porous and tortuous structure of the regenerator requires careful investigation. In the context of thermoacoustics, the matter is difficult to address not only because the oscillatory nature of the flow but also because the correctly defined phasing between pressure and velocity needs to be considered.

However, most of the past experiments have not considered specific phase angles between pressure and velocity oscillations. Therefore, their effect on the regenerator hydrodynamic condition is still unknown. This necessitates the development of an appropriate closure model for a regenerator working in the travelling-wave mode since a hydrodynamic correlation based on an unspecified phase angle may introduce unknown errors. Since the regenerators are porous media they can be studied theoretically on the grounds of the porous medium theory [4]. This allows modelling a porous structure through the use of additional parameters in the transport equation that represent the resistance to the flow.

Until now, the behaviour of the oscillatory flow across a porous medium is not fully understood. The matter is complicated further by the requirement of a specific phase angle between pressure and velocity for the thermoacoustic device to function properly. In addition, the lack of a well-defined friction correlation for oscillatory flow across porous medium complicates the determination of correct closure model for the purpose of numerical modelling. These issues are addressed in the current work by the application of a mixture of experimental and CFD approaches.

2. Literature Review

The friction factor is a dimensionless parameter used to represent the hydrodynamic condition of a flow across a structure. In oscillatory flow the friction factor is defined as [5]:

$$f_{osc} = \frac{X_{\Delta P} d_h}{\frac{1}{2}\rho X_{u_m}^2 L_r}. \tag{1}$$

The terms $X_{\Delta P}$, d_h, ρ, X_{u_m}, and L_r represent the amplitude of pressure drop, hydraulic diameter, gas density, amplitude of velocity and length of the regenerator, respectively. The subscript u_m relates to the mean value taken between two measured points. The hydraulic diameter, d_h, is defined as [5]:

$$d_h = \frac{\phi}{1-\phi} d_w. \tag{2}$$

The value of porosity, ϕ, and the wire diameter, d_w, are usually supplied by the manufacturer of mesh screens. The experimentally measured pressure drop and velocity are used to calculate the friction factor. The results calculated through Equation (1) may be fitted with suitable correlation to represent the hydrodynamic condition of the flow investigated. The basic form of friction factor correlation is given by the standard two-parameter Ergun equation [6]:

$$f = \frac{a_1}{Re} + a_2, \tag{3}$$

where a_1 and a_2 are constants that fit the condition tested. Swift and Ward [7] showed that the two constants of Equation (3) can be represented by an equation for porosity defined as:

$$a_1 = 1268 - 3545\phi + 2544\phi^2, \tag{4}$$

$$a_2 = -2.82 + 10.7\phi - 8.6\phi^2. \tag{5}$$

In their correlation, Swift and Ward [7] defined the Reynolds number by taking into consideration the simple harmonic oscillation in regenerator and expressed it as the first order amplitude of complex Reynolds number, Re_d, given by:

$$Re_d = \frac{4 < u_1 > r_h \rho_m}{\mu}, \tag{6}$$

where $< u_1 >$, r_h, ρ_m and μ are the first order spatial average velocity, hydraulic radius, mean density and viscosity, respectively. Assuming a simple harmonic oscillation in the regenerator, their correlation based on Equations (3)–(5) is reported to fit well the steady experimental data of Kays and London [8]. This equation had been widely used in thermoacoustic community in conjunction with DeltaEC, design software for thermoacoustic devices [9]. The suitability of Swift and Ward's equation for use in an oscillatory flow is however questionable because it is fitted to steady flow data.

A number of research works considered various specific formulations of equations for friction factor, including Gedeon and Wood [6], Ju et al. [10], Nam and Jeong [11] or Boroujerdi and Esmaeili [12]. These formulations are generally of the same form as the standard Ergun Equation (3), but are typically multiplied by a certain dimensionless modification factor. This is typically obtained through dimensional analysis, experiments, numerical simulations or their useful combination, and its presence highlights the important differences between oscillatory and steady flow conditions.

Another important parameter in an oscillatory flow is referred to as (acoustic) impedance. It is defined as a ratio between pressure and velocity of a flow. However, impedance is a complex number and so in addition to their amplitudes it also includes the phase difference between them [3]. In travelling wave systems, the phase difference between pressure and velocity can be controlled by an RLC network which is also known in some papers as a "phase shifter". Liu et al. [13] reported that the phase shifter reduces losses within the regenerator and improves system performance. This indicates that the pressure drop characteristics within the regenerator depend on the phase shift between pressure and velocity. In most of the papers discussed above, the phase difference between pressure and velocity is not set experimentally to a specific value. This may introduce additional unknown effects. There is yet no clear explanation as to how this affects the friction losses of the whole system. It is found, however, that several situations have been reported in the literature where the meaning of friction factor is unclear when the phase difference is present.

Hsu [14] showed that when a phase difference exists, setting the amplitude alone is not sufficient to predict the hydrodynamics of flow in the porous medium. The experimental detail is reported by Hsu et al. [15]. The coefficients a_1 and a_2 of the friction correlation (Equation (3)) are presented by Hsu et al. [15] as Darcy and Forchheimer coefficients. The friction factor gained from experimental results is discussed from a theoretical point of view. The theoretical explanation involves Stokes drag force, C_s, frictional force due to boundary layer, C_B, and inviscid form drag, C_I, first defined by Hsu and Cheng [16] and further developed theoretically by Hsu [14]. The theoretical prediction in that study is in good agreement with experimental results when the effect of phase difference is considered. In this paper, the phase difference between pressure and velocity will be referred to as "phase shifting" for brevity.

The findings of Hsu [14] may be related to the "phase shifting" phenomena reported by Ju et al. [10] and the breathing factor, B, introduced by Nam and Jeong [11]. The phase shift effect is also observed in the experimental study of Zhao and Cheng [17] when the kinetic Reynolds number increases with frequency. Clearly the phase shifting between pressure and velocity has an effect on friction losses reported in many investigations of oscillatory flows. There is a possibility that these phenomena occur as a result of the experimental conditions not being set to a particular time-phasing between pressure and velocity. However, no attention has been given to the possibility of controlling the time-phase between pressure and velocity in the experimentation for determination of the hydrodynamic condition of the oscillating flow through a regenerator. This is crucial in thermoacoustic applications because the regenerator should work with a well-controlled travelling-wave time-phasing.

The porous and tortuous structure of the regenerator is difficult and computationally very expensive to model using CFD if the full original structure is to be replicated. An economical way of modelling such a structure is through the use of the porous medium theory.

The porous medium theory models a porous structure through the use of additional parameters in the transport equation that represent the flow resistivity. Flow resistance is modelled through the knowledge of pressure drop caused by the structure. According to the Darcy–Forchheimer model, the pressure gradient, ∇p, is defined as [4]:

$$\nabla p = -\frac{\mu}{K}\mathbf{v} - FK^{-1/2}\rho|\mathbf{v}|\mathbf{v}, \tag{7}$$

where μ, ρ, and $\mathbf{v}$ are the gas dynamic viscosity, density and velocity, respectively. The terms K and F represent the additional parameters representing form drag to the flow and the terms are known as permeability coefficient and Forchheimer inertial coefficient, respectively. In the simplest case, the flow across the porous medium can be represented by Darcy's law. Darcy's law simply neglects the second term on the right hand side of Equation (7) leaving only the term with permeability coefficient, K, to represent the porosity experienced by the flow. This law is valid only when the Reynolds number, defined by the average flow velocity, is an order of magnitude smaller than one. When the velocity is high, the Forchheimer modified equation should be considered to account for the inertial losses in the porous medium. In this condition, Equation (7) should be applied in full. In a highly viscous flow, a further modification is suggested where the second term on the right hand side of Equation (7) is replaced with the Brinkman's term, $K\nabla^2\mathbf{v}$.

Darcy's permeability, K, and Forchheimer inertial coefficient, F, are empirical constants. As empirical constants, both permeability and inertial coefficient are uniquely dependent on the porosity and tortuousity of the porous structure (regenerator). The values determine the momentum losses occurring in the porous media. In experimental practice, the momentum loss can be obtained from the pressure drop measured across the regenerator. On the other hand, to model a system with regenerator, one will need to determine the permeability, K, and Forchheimer inertial coefficient, F, beforehand. This becomes a challenge especially due to the lack of friction factor correlations suitable for oscillating flows in porous media.

Cha et al. [18] reported a CFD-assisted method for determining the permeability and inertial coefficient of a porous regenerator in an oscillating flow. The regenerator in their study, limited to a few sizes of mesh screens, is tested in a small sized device where the inertial effect could be significant. The oscillating flow friction factor is shown to deviate from the steady flow data of Clearman et al. [19] when the permeability Reynolds number, $Re_K = \rho u K^{1/2}/\mu$, is greater than 0.1. Landrum et al. [20] followed the same CFD procedure in establishing the hydrodynamic properties of the regenerator in an oscillating flow but with small mesh fillers suitable for miniature cryocoolers. In a different study, the porous medium coefficients were also derived by Tao et al. [21]. However, the derived equation is inappropriate for general use as the friction factor applied in defining the pressure drop is the equation specifically built for cryogenic conditions.

3. Experimental Setup

The schematic diagram of the test rig is shown in Figure 2. In essence, it is a travelling-wave cooler in which a sample of regenerator material can be placed and the resulting pressure drop is measured. The rig consists of a linear motor, resonator, hot and cold heat exchanger, regenerator (mesh numbers #30, #94, #180, and #200 have been used) and a set of components known as resistance, R, inertance, L, and compliance, C (forming an RLC network). The network is a combination of suitable valves, a 1.6 m long tube with a diameter of 8 mm and a buffer volume of 2.5×10^{-3} m^3. The frequency of the acoustic wave is set using a linear motor. The linear motor is enclosed in a specially designed high-pressure cylindrical casing with a glass window to allow laser displacement measurement of the motor piston. The laser displacement sensor (Keyence LK-G152, Milton Keynes, UK) senses the piston displacement, δ. The amplitude of the velocity at that location, defined as V_1, is then calculated as $V_1 = \omega\delta$.

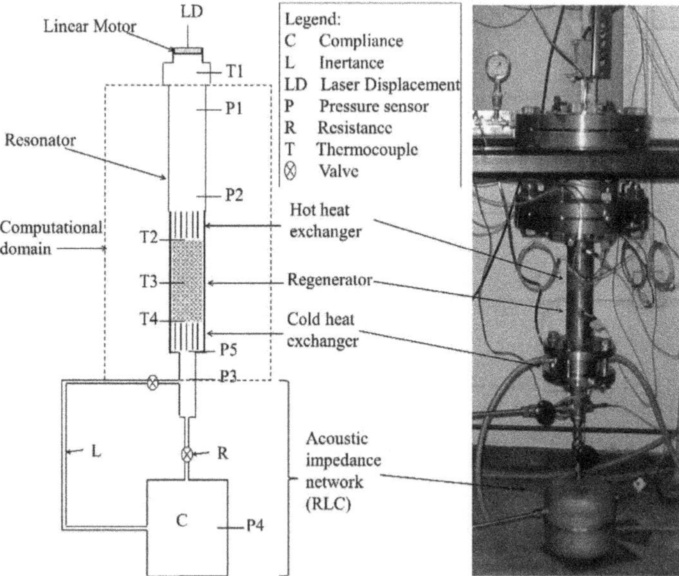

Figure 2. Schematic representation of the experimental rig and the selected computational domain.

The resonator has the inner diameter of 55 mm. The hot heat exchanger, regenerator and cold heat exchanger are made to fit into the resonator. The lengths of the resonator, hot heat exchanger, regenerator and cold heat exchanger are 185 mm, 25 mm, 50 mm and 27 mm, respectively. The regenerator is built using stainless steel mesh screens stacked on top of one another to fill up the 50 mm length. The supplied acoustic wave creates a temperature gradient along the regenerator. The development of the temperature gradient is controlled to concentrate only on the hydrodynamic condition of the regenerator. This is achieved through the cold heat exchanger located at the bottom end of the regenerator. The cold heat exchanger is connected to a reservoir that supplies water at room temperature so as it is kept at room temperature. The temperatures at locations T1, T2, T3 and T4 are measured by type-K thermocouples. The rig typically operates with helium at mean pressure up to 60 bar.

3.1. Data Collection and Reduction

The phase angle between pressure and velocity is controlled via the RLC network as shown in Figure 2. The valves are adjusted accordingly until the travelling-wave condition is achieved. The phase is observed by monitoring the phase angle, φ, of impedance at location 5. The impedance at location 5, Z_5, is given as:

$$Z_5 = \frac{P_5}{u_5} = |Z_5|(\cos\varphi + i\sin\varphi). \tag{8}$$

The terms P_5, u_5, and φ represent the oscillating pressure, oscillating velocity and phase angle of impedance at location P5, respectively. The phase angle of impedance represents the phase difference between pressure and velocity at that location. When the phase angle of impedance is zero, the imaginary part of impedance becomes zero. Hence oscillating pressure and oscillating velocity are both contributing only to the real value of impedance. As a result, pressure and velocity are in phase [22]. Pressures are measured in the experiment. The velocity at location 5, u_5, is determined using a Transfer Matrix Method. The Transfer Matrix Method has originally been used to measure the acoustical properties of components such as the porous media in the study by Song and Bolton [23] and Ueda et al. [24]. Ueda et al. [24] applied a linear thermoacoustic theory to estimate acoustic characteristics of a regenerator. In the current study, the method is used to theoretically provide information on variables not measured in the experiment. Lossless thermoacoustic model is used to relate pressure, P, and velocity, u, at two locations. This is shown in a matrix form as follows:

$$\begin{bmatrix} \frac{\partial P}{\partial x} \\ \frac{\partial u}{\partial x} \end{bmatrix} = \begin{bmatrix} 0 & -i\omega\rho_m \\ \frac{i\omega}{\gamma p_m} & 0 \end{bmatrix} \begin{bmatrix} P \\ u \end{bmatrix}. \tag{9}$$

These lossless equations are based on assumptions that the flow is adiabatic and inviscid. The assumptions are considered valid between points 3 and 5 due to the short distance between them and the absence of the influence of temperature within the area (as indicated in Figure 2). The solutions of Equation (9) may be shown in a matrix form as:

$$\begin{bmatrix} P_5 \\ u_5 \end{bmatrix} = \begin{bmatrix} m_{11} & m_{12} \\ m_{21} & m_{22} \end{bmatrix} \begin{bmatrix} P_3 \\ u_3 \end{bmatrix}. \tag{10}$$

Four equations are needed to solve for the four unknowns. Two equations are obtained from the lossless thermoacoustic model with analytical solutions (as given by Swift [25]) cast into the matrix form of Equation (10) to represent m_{11} and m_{21} and are shown as follows:

$$m_{11} = \cos(kx), \tag{11}$$

$$m_{21} = \frac{i}{\rho a}\sin(kx). \tag{12}$$

Two more equations are obtained through the wave characteristics following the analysis of Song and Bolton [23]. The flow is reciprocal and the area of interest is symmetrical. For symmetrical system, m_{11} equals m_{22}. The reciprocity of the flow requires that the determinant of the transfer matrix be unity. These two characteristics are shown as follows:

$$m_{11} = m_{22}, \tag{13}$$

$$m_{11}m_{22} - m_{12}m_{21} = 1. \tag{14}$$

The transfer matrix method used in the experiment is finally shown as:

$$\begin{bmatrix} P_5 \\ u_5 \end{bmatrix} = \begin{bmatrix} \cos(kx) & i\rho a \sin(kx) \\ \frac{i}{\rho a}\sin(kx) & \cos(kx) \end{bmatrix} \begin{bmatrix} P_3 \\ u_3 \end{bmatrix}. \tag{15}$$

The terms ω, ρ, and a are the angular velocity, density and sound speed, respectively. The velocity at location P_3, u_3, is calculated using the lumped element method with the equation given by Swift [25]:

$$u_3 = \frac{i\omega V_c P_4}{S\gamma p_m}. \tag{16}$$

In the lumped element method, the velocity is calculated using information gained from the compliance. The volume of the compliance, V_c, and its cross-sectional area, S, are measured manually, while pressure, P_4, is measured by the pressure transducer located at the compliance. p_m is the mean pressure set for the experiment.

Four differential pressure transducers (PCB#112A21), connected to a signal conditioner (PCB Piezotronics 480B21, Depew, NY, USA), are used to measure the amplitude of the oscillating pressure at locations P1, P2, P3, P4 and P5 (refer to Figure 2). Data is collected using a PC-based data acquisition board. The phase angle is monitored for each test condition to ensure that a near travelling-wave phasing is achieved in every test.

The experimental results are used to calculate the friction factor using Equation (1). The velocity used for calculating the friction correlation of Equation (1) is an averaged value calculated as:

$$u_m = \frac{u_1 + u_5}{2\phi}, \tag{17}$$

where u_1 and u_5 are the amplitudes of velocities at locations LD and P5, respectively. Strictly speaking, velocity u_2 should be included in Equation (17), but this cannot be measured directly. Fortunately, there is only a short distance of empty resonator between locations 1 and 2, while the wave is travelling type. According to early design estimates $u_2 \approx u_1$ to within 1%, while u_1 is easy to measure directly with the laser displacement sensor. The velocity, u_m, represents the actual velocity amplitude in the regenerator. All the regenerator samples are tested at a frequency of 30 Hz and a mean pressure of 25 bar.

3.2. Properties of Gas and Dimensions of Solid Matrix

Four different sizes of mesh screens were tested. The dimensions and parameters of each regenerator mesh screen are listed in Table 1. The working medium is helium, commonly used in thermoacoustic applications because of its good thermal conductivity and specific heat. This allows producing large temperature gradients for a better heat pumping effect. The mesh screen is made of stainless steel. The properties of both the gas and solid material are shown in Table 2. These properties are obtained at the temperature of 300 K.

Table 1. Geometric dimensions of regenerators tested.

Mesh Screen Regenerator (Material: Stainless Steel)				
Regenerator	**Mesh Number**	**Wire Diameter (mm)**	**Porosity (%)**	**Hydraulic Radius (μm)**
1	180	0.058	67.5	30.31
2	200	0.041	74.8	30.27
3	30	0.28	72.7	195
4	94	0.089	74.2	63.79

Table 2. Properties of helium gas and stainless steel solid matrix.

Properties	Helium	Stainless Steel
Ratio of heat coefficient, γ	1.6667	-
Speed of sound, a (m/s)	1019.2	-
Density, ρ (kg/m^3)	4.0115	7918.1
Specific heat, c_p (J/kg·K)	5193.4	453.84
Gas expansion, β (1/K)	3.3333×10^{-3}	-
Thermal conductivity, k (W/m·K)	0.15243	14.388
Dynamic viscosity, μ (kg/m·s)	1.9938×10^{-5}	-

4. Computational Model

The regenerator is modelled using a porous medium theory. Two-dimensional Darcy–Forchheimer porous medium equations are expressed as follows [14]:

$$\phi \frac{\partial \rho}{\partial t} + \nabla \cdot (\rho \mathbf{v}) = 0, \tag{18}$$

$$\frac{\partial (\rho \mathbf{v})}{\partial t} + \nabla \cdot \left(\frac{\rho \mathbf{v} \cdot \mathbf{v}}{\phi} \right) + \nabla p - \nabla (\boldsymbol{\tau}) = -\left(\frac{\mu \phi}{K} \mathbf{v} + \frac{F \phi^2 \rho}{\sqrt{K}} \mathbf{v} |\mathbf{v}| \right). \tag{19}$$

The terms ϕ, ρ, t, $\mathbf{v}$, p, $\boldsymbol{\tau}$, μ, K and F represent the porosity, fluid density, time, velocity vector, porous pressure, stress tensor, fluid dynamic viscosity, permeability coefficient and Forchheimer inertial coefficient, respectively. For a homogeneous porous medium with local thermal equilibrium assumption, the one-dimensional energy equation for this porous region is given as [4]:

$$\begin{aligned} &\left[\left(\phi k_f + (1-\phi) k_s \right) \nabla^2 T + (1-\phi) q_s''' + \mathbf{v} \left(-\tfrac{\partial P}{\partial x} + \rho_f g \right) \right] \\ &= \left(\phi \rho_f c_{pf} + (1-\phi) \rho_s c_s \right) \tfrac{\partial T}{\partial t} + \rho_f c_{pf} \mathbf{v} \tfrac{\partial T}{\partial x} \end{aligned} \tag{20}$$

where q_s''' refers to the heat generation rate and T is the temperature. The term c_p is the isobaric specific heat and c_s is the specific heat. The subscripts f and s refer to fluid and solid constituents, respectively. Assuming local thermal equilibrium, the temperature of the gas and solid structure is the same at any time and spatial location due to the condition of perfect thermal contact of the regenerator.

In ANSYS-FLUENT [26], the porous medium is solved using equations given as:

$$\frac{\partial (\phi \rho)}{\partial t} + \nabla \cdot (\phi \rho \mathbf{v}) = 0, \tag{21}$$

$$\frac{\partial (\phi \rho \mathbf{v})}{\partial t} + \nabla \cdot (\phi \rho \mathbf{v} \mathbf{v}) + \phi \nabla p + \nabla \cdot (\phi \tau) = \phi B_f - \left(D_x \mu + \frac{C_x \rho}{2} |\mathbf{v}| \right) \mathbf{v}, \tag{22}$$

$$\frac{\partial}{\partial t} \left(\phi \rho_f E_f + (1-\phi) \rho_s E_s \right) + \nabla \cdot \left(\mathbf{v} \left(\rho_f E_f + p \right) \right) = \nabla \cdot \left[k_{eff} \nabla T - (\boldsymbol{\tau} \cdot \mathbf{v}) \right]. \tag{23}$$

The term $E_f = c_p T - P/\rho + \mathbf{v}^2/2$ is the energy of the fluid and $k_{eff} = \phi k_f + (1-\phi) k_s$ is the effective thermal conductivity within the porous regions. Note that the velocity vector $\mathbf{v}$ used in ANSYS FLUENT [26] corresponds to the superficial velocity. The real value of velocity within the porous region is obtained by dividing the superficial velocity by the porosity of the region, ϕ. Relationships between coefficients C_x and D_x as used in [26] and the coefficients of the standard porous medium theory (permeability coefficient, K, and Forchheimer inertial coefficient, F) may be obtained by comparing Equations (19) and (22). Note that the viscous dissipation represented by the stress tensor, $\boldsymbol{\tau}$, and the body force, B_f, in Equation (22) are neglected. The relationships obtained are shown as follows:

$$K = \frac{\phi^2}{D_x}, \tag{24}$$

$$F = \frac{C_x\sqrt{K}}{2\phi^3}. \tag{25}$$

These coefficients characterise the hydrodynamic conditions of a porous medium that represent momentum sink (pressure drop) caused by the presence of the medium. As reviewed earlier, the Darcy model suggested that the permeability coefficient alone is enough to represent the pressure drop in a low speed flow. As the speed increases, an inertial effect takes place and the Forchheimer coefficient becomes important to improve the flow model.

4.1. Computational Domain

Figure 3 presents the two-dimensional axis-symmetrical computational domain used in this study. Flow in the domain occupied by the regenerator is solved using the porous medium theory. Elsewhere, the flow and energy transfer are solved using standard unsteady Navier–Stokes equations available in [26].

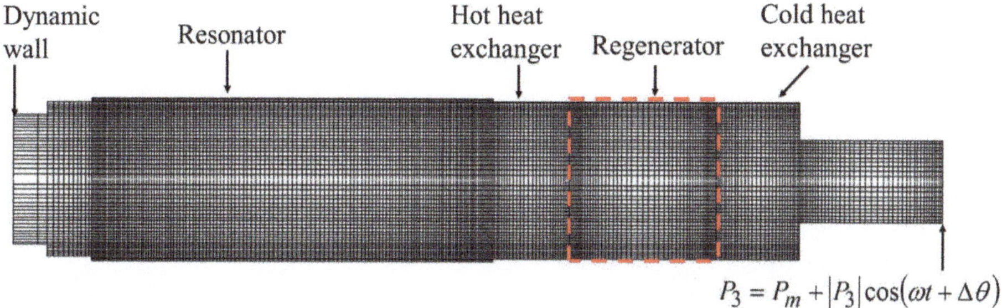

Figure 3. Computational domain for the regenerator study. The red-dashed box shows the regenerator domain (porous medium).

A local thermal equilibrium is assumed for the energy equation. Hydrodynamic losses may be attributed to the viscous shear (skin friction) due to the presence of the wall, and a pressure force (form drag) due to the presence of a solid body [8]. Generally, both effects are considered as a package that contributes to resistance losses. However, ANSYS FLUENT [26] defines body forces as separate from the momentum sink that represents the porous medium. Structures investigated in this study have a relatively high porosity. Furthermore, the flow investigated is limited to a low speed. For the conditions of the current study, it is found that the highest pressure drop due to the body force of the porous structure investigated is less than 0.1%, and hence is neglected.

In this study, the porous medium has been treated as isotropic. The losses in the direction other than the main flow direction were assumed small and negligible. The inertial effect was neglected because of the low speed of the flow. Pressure and velocity were expressed in terms of mean components and first order fluctuating components (except that the mean velocity is zero). The simple harmonic equations (cf. Figure 3) were applied in defining the boundary conditions of the computational domain.

A dynamic mesh was imposed as the inlet boundary condition to replicate the movement of the piston of the linear motor. The dynamic mesh area and the resonator are separated by a short tube (15 mm in length and 50 mm in diameter). The computational domain excluded the RLC network, but imposed the phase control using a user-defined-function (UDF) at the inlet wall and the exit pressure at location P3 (refer to Figures 2 and 3). The last region after the cold heat exchanger is 75.5 mm long and has a diameter of 27.17 mm. The working gas was treated as a compressible ideal gas. The model has been solved using an unsteady pressure-based implicit solver with the first order implicit scheme for the discretisation of time. For model that uses dynamic mesh, the time discretisation scheme is only available in ANSYS as first order calculation [26]. Therefore, first order calculation with implicit scheme was used. The time step size was set to 1.11111×10^{-4} s (which corresponds to 300 steps per acoustic cycle at frequency of 30 Hz) so that convergence could be achieved within 15–18 iterations per time step. The convergence was set to 10^{-4} for continuity and momentum equations and 10^{-6} for energy equation. Pressure–velocity coupling has been solved using pressure-implicit-with-splitting-operators (PISO) algorithm available in ANSYS FLUENT [26]. All variables (pressure, density, momentum and energy) were discretised using second-order upwind method. The model was run until steady oscillatory flow condition was achieved. The steady oscillatory condition was monitored through the time history data of velocity inside the resonator. It was observed that the model needs to be run for at least five cycles in order to achieve the steady oscillatory flow condition.

The operating pressure was set at 25 bar. The flow has a frequency of 30 Hz. Temperatures at the location of the hot and cold heat exchangers were set as constants with values following the measured temperatures in the experiment. The temperature developed reflects the result of the thermoacoustic effect occurring at the regenerator when the travelling-wave time phasing is set. The temperature gradient within the regenerator, for all mesh screen samples, varies depending on the acoustic excitation and type of mesh screens used. In an attempt to minimize the effect of temperature on the pressure readings, the development of temperature gradient has been controlled using cold heat exchanger. For all regenerator samples tested here, the recorded temperature gradient varies in a range of $10\ ^{\circ}\mathrm{C} \leq \Delta T \leq 21\ ^{\circ}\mathrm{C}$. The values of temperature from experiments are used in the model so that the effect of the small temperature variations on pressure readings are being modelled following the experimental findings.

The contribution of pressure loss from the heat exchanger has been calculated using the pressure drop prediction for a compact heat exchanger as presented in [8]:

$$\frac{\Delta p}{p_1} = \frac{v_1^2/2g}{p_1/\rho_1}\left[\left(K_c + 1 - \phi^2\right) + 2\left(\frac{1}{\rho} - 1\right) + f\frac{L}{r_h}\rho - \frac{1}{\rho}\left(1 - \phi^2 - K_e\right)\right]. \tag{26}$$

The subscript 1 refers to parameters obtained at the inlet of the heat exchanger. The terms p, v, ρ, g, ϕ, and r_h refer to pressure, velocity, fluid density, gravity, porosity and hydraulic radius, respectively. Parameters K_c, K_e and f are the entrance losses, exit losses and Fanning friction losses, respectively, as given in [8]. The calculations showed that the contribution of cold heat exchanger losses was within 1%–5% of the whole pressure drop measured between locations P2 and P5. The loss varies according to the range of velocity and pressure drop achieved in the experiment. The overall contribution of losses is small because of the low speed of flow. The calculated loss was deducted from the total pressure measured in the experiment. The hydrodynamic condition of the regenerator for CFD modelling was gained by determining the permeability coefficient that gives a pressure drop similar to the experimentally measured value.

4.2. Pressure Drop and Grid Size

The mesh was defined to be denser near the wall and at the interfaces between the rig components inside the resonator. The mesh near the dynamic wall was made coarser to allow for the movement of the wall. The grid independency test has been carried out by increasing the number of mesh points by a factor of 1.3. The grid structure remained unchanged with only the mesh density being increased. The resulting pressure drop obtained for grids of different sizes is shown in Figure 4. A grid size with a total number of cells of 8794 has been found to be sufficient to provide a solution that is independent of the grid and therefore selected for this investigation. The model has also been run in both single and double precision solver in readiness to test for round-off error. It appears that the single precision is sufficient to solve this two-dimensional axis-symmetrical model.

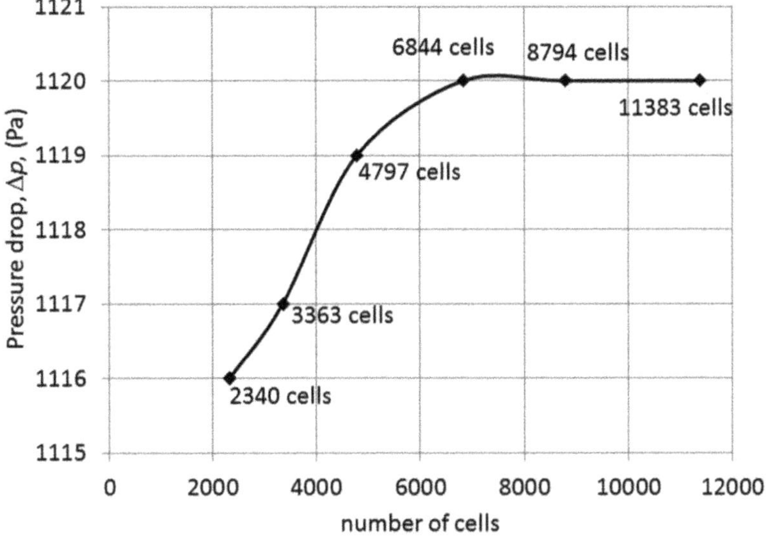

Figure 4. Grid independency test for pressure drop in regenerator made of mesh screen #200.

4.3. CFD Model Validation

The model was pre-validated for boundary conditions and subsequently the porous coefficients C_x and D_x were set followed by the final validation of the porous model. In the pre-validation stage, only the velocity and pressure at the location before the porous medium (locations P1 and P2 in Figure 2) were compared to the experiment. A good agreement was found as shown in Figure 5.

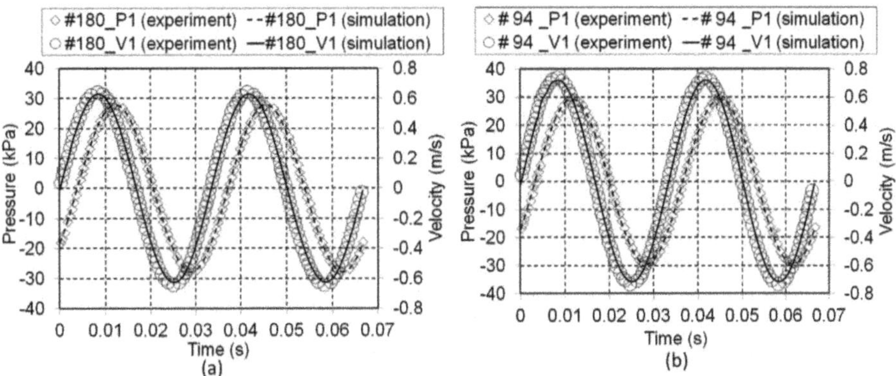

Figure 5. Instantaneous pressure at P1 and velocity at the piston for: (**a**) #180; and (**b**) #94.

Once the model had been pre-validated, the permeability coefficient, K, of the regenerator was then predicted by carefully increasing the permeability value in ANSYS FLUENT [26] until the pressure drop matched the experimental value. As soon as the pressure drop matched, the validation was finalised by comparing the inlet velocity, calculated from the measured displacement of linear motor's piston (V1 in Figure 5), and pressures at locations P2 and P5 (definition of locations is given in Figure 2). The same procedure was repeated for each case until a good agreement with the experiment was obtained for all samples tested. Figure 6 shows the results for pressure at locations P2 and P5 after the permeability tensor set matched the pressure drop measured in the experiment.

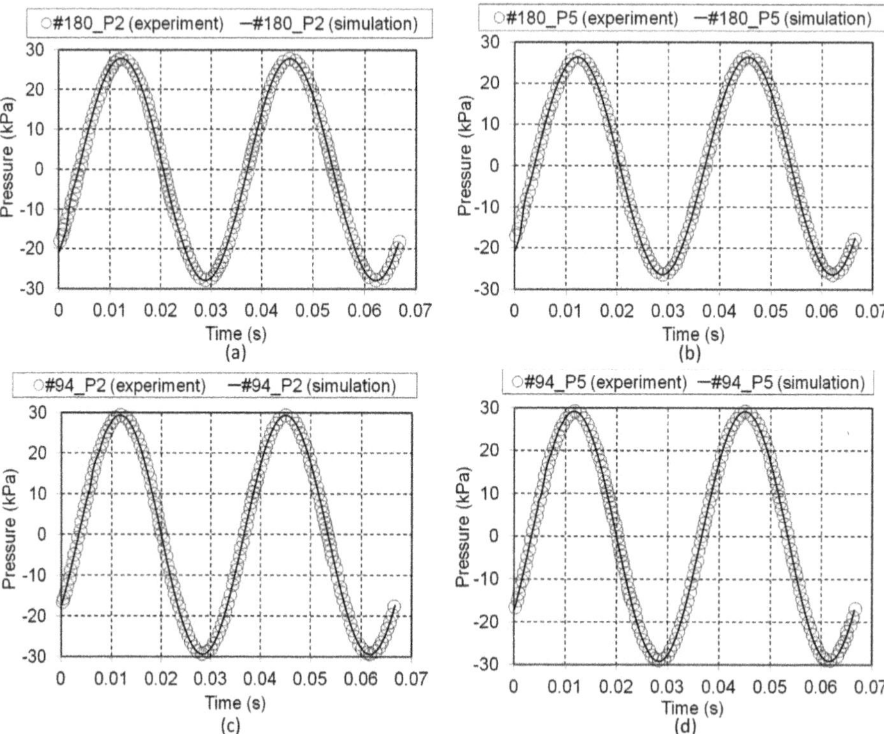

Figure 6. Validation of instantaneous pressures at locations P2 and P5: (**a**,**b**) for #180; and (**c**,**d**) for #94.

5. Results and Discussion

The hydrodynamic conditions of the regenerator samples tested are presented as friction factor. The friction factor was derived from the Darcy model and results are presented using the permeability coefficients obtained from the numerical work. Comparison to experimental results and Swift and Ward correlation [7] is also made. Further analysis on the phase angle between pressure and velocity is also shown.

5.1. Friction Factor Correlation

In the porous medium theory, the friction factor of most porous media is presented by the permeability Reynolds number, $Re_K = \rho u K^{1/2}/\mu$. The pressure drop characteristics of the porous medium are related to the permeability Reynolds number as given by [4]:

$$-\nabla P = \frac{1}{2}\frac{f\rho u^2}{\sqrt{K}}. \tag{27}$$

The pressure gradient can also be represented by the Darcy–Forchheimer model as in Equation (7). The following formula can be derived from Equations (7) and (27):

$$f = \frac{2\sqrt{K}}{\rho u^2}\left[\frac{\mu}{K}u + \frac{F\rho}{\sqrt{K}}u^2\right]. \tag{28}$$

Further modification of Equation (28) can be made to include the hydraulic diameter as follows:

$$f = \frac{2\sqrt{K}}{\rho u^2} \cdot \frac{d_h}{d_h} \left[\frac{A_1 \mu}{K} u + \frac{A_2 F \rho}{\sqrt{K}} u^2 \right], \tag{29}$$

$$f = \frac{2d_h}{\sqrt{K}} \cdot \frac{A_1}{Re_h} + 2A_2 F. \tag{30}$$

The permeability coefficient, K has a unit (length)^2 and the Forchheimer inertial coefficient, F, is dimensionless. Parameters A_1 and A_2 are constants introduced to correlate the equation to fit the experimental result. Simple dimensional analysis shows that Equation (30) is dimensionless. The derived correlation agrees with the well-known Ergun form presented in Equation (3). From this derived friction factor, the coefficients a_1 and a_2 in the Ergun equation are defined as:

$$a_1 = \frac{2A_1 d_h}{\sqrt{K}}, \tag{31}$$

$$a_2 = 2A_2 F. \tag{32}$$

In the simplest Darcy model, when the velocity is small, the inertia effect is neglected and the friction factor is left with only first term of right-hand side of Equation (30):

$$f = \frac{2d_h}{\sqrt{K}} \cdot \frac{A_1}{Re_d}. \tag{33}$$

Finally, with this correlation the porous coefficient predicted through CFD model is used to predict the friction factor of the porous medium and then compared to the experimentally calculated friction factor for the actual sample. The friction factor correlation presented in Equation (33) is derived from the Darcy correlation. Hence comparing the simulation results to experiment requires the friction factor predicted by simulation model to be divided by four. In this study, all comparisons are carried out in line with the Fanning-based correlation. The Fanning-based correlation for computational models is given as:

$$f = \frac{d_h}{2\sqrt{K}} \cdot \frac{A_1}{Re_d}. \tag{34}$$

Permeability is an empirical constant used in the porous medium community to characterize the ability to transmit fluid by the porous media. The permeability coefficients for the regenerator samples investigated, obtained from the CFD-assisted method, are tabulated in Table 3.

Table 3. CFD-predicted permeability for regenerator investigated.

Regenerator Mesh Screen Numbers	#200	#180	#94	#30
Permeability, K (m^2)	1.1505×10^{-10}	0.99012×10^{-10}	10.5715×10^{-10}	41.29×10^{-10}

A higher regenerator mesh number refers to a finer mesh. This is reflected by the value of the wire diameter of each sample as listed in Table 1. The permeability is high for a regenerator with a mesh screen number of #30 and decreases as the regenerator mesh screen number increases to #200 (the value for #180 being slightly outside this trend due to a lower porosity value—cf. Table 1). The permeability coefficient in Table 3 is used to calculate the friction factor using Equation (34). Several cases of measurement and numerical prediction were considered for each sample. These corresponded to different level of acoustic excitation produced by the linear motor, and result in different values of Reynolds number, Re_d, as defined in Equation (6). The results are plotted in Figure 7.

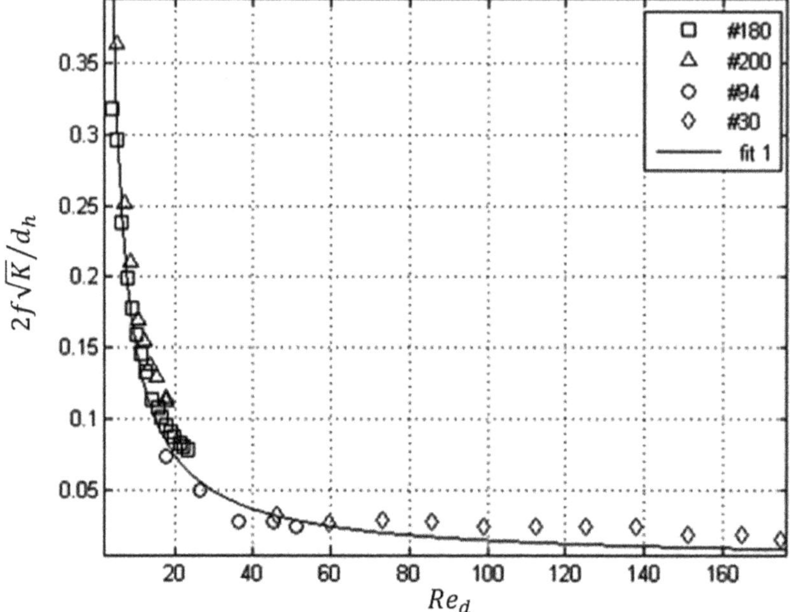

Figure 7. Correlation of friction factor in terms of d_h, K and Re_d.

In Figure 7, there are 17 square, nine triangle, six circle and 11 diamond symbols that represent tests of samples #180, #200, #94 and #30, respectively. A fit can be obtained for all the cases as shown by the line in the figure. The friction data obtained from the numerical model fitted the following correlation:

$$f = \frac{d_h}{2\sqrt{K}} \cdot \frac{1.48}{Re_d}. \tag{35}$$

The correlation was obtained based on the least-square fit with a confidence level of 95%. The correlation represents a friction loss of all the four regenerator mesh screens tested. The correlation best represents the hydrodynamic losses of high regenerator mesh numbers and slightly under-predicts the losses occurring at a regenerator build with low mesh screen numbers. The maximum difference between the correlation and the friction value of regenerator with low mesh screen number are 30% and 50% for #94 and #30, respectively. The diamond symbols for Re_d larger than 60 are clearly deviating from the line. This may be due to the fact that for large Reynolds number and a very coarse size of regenerator mesh screens the Forchheimer inertial coefficient may become significant and Equation (30) may apply. The constant 2 in the denominator of Equation (35) is the result of the conversion from Darcy-based correlation to Fanning-based correlation and left on purpose for clarity.

Figure 8 shows that the numerical results represented by correlation (35) and the results calculated using Swift and Ward [7] correlation (cf. Equations (3)–(5)) both qualitatively agree with experimental measurement. Swift and Ward [7] defined their friction correlation using Darcy-based correlation, and therefore the results calculated through Equation (3) through to Equation (5) need to be divided by 4 for comparison. The results indicate that the assumption used by Swift and Ward [7] is acceptable. The assumption states that pressure drop and velocity relate in a similar way as in steady flows [7]. There are arguments found in the literature that claimed that a pressure drop condition in oscillatory flow deviated from predictions of steady flow. An explanation to that situation could relate to the influence of phase-shifting between pressure and velocity. Further discussion of this is provided in Section 5.2 for low mesh number regenerators. Current results have shown that a pressure drop condition for an oscillatory flow with travelling-wave time-phasing does not deviate much from the Swift and Ward's correlation, especially when the mesh number is high.

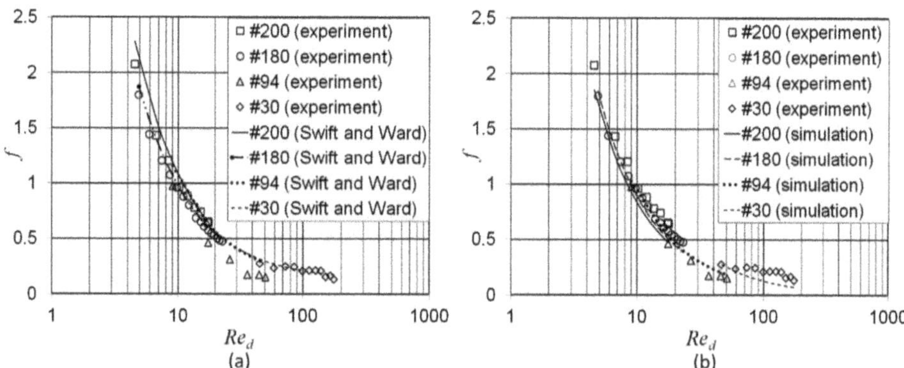

Figure 8. Comparison between friction factor from experiment: (**a**) Swift and Ward correlation of [7]; and (**b**) current CFD-assisted correlation.

5.2. Analysis of Phase Shift

Phase angle between pressure and velocity oscillations is important in determining the hydrodynamic condition of porous media such as the regenerators. If the pressure and velocity are not in phase, the maximum pressure occurs at a certain phase lag compared to maximum velocity. Then, the amplitude data may not be accurate enough to represent the friction factor according to Equation (1).

In other studies [10,11,18], the phase shift between pressures at location before and after the regenerator is observed when the operating frequency increases. The phase shift effect is observed even at a frequency as low as 5 Hz in the results presented by Cha et al. [18]. Zhao and Cheng [17] reported that the phase shift between pressure and velocity in the experimental investigation appeared at a very low kinetic Reynolds number (calculated to be equivalent to 2.5 Hz).

In the current experiment, it is not possible to measure the pressure drop and velocity amplitude within the regenerator. However, the numerical model is validated for several points within the rig. Thus, it is reasonable to expect that it will reliably predict the pressure drop and velocity within the regenerator.

Figure 9 shows the pressure drop and velocity within the regenerator. The pressure drop was calculated using values obtained at locations P2 and P5, where pressure sensors are available. Note that vertical scales of the plots in Figure 9 vary. This was done deliberately to show more clearly the changes in phase difference between pressure and velocity. For fine regenerator mesh screens (#200 and #180), the pressure drop is in phase with the velocity within the regenerator. When the mesh screen is coarser (#30 and #94), the pressure drop becomes slightly out of phase from the velocity within the regenerator. The pressure drop tends to lead the velocity more visibly as the mesh becomes coarser. Theoretically, bigger porosity results in a lower flow velocity within the porous region, if set at a similar level of acoustic excitation. The shear stress and form drag is likely to be smaller too. However, due to the limitation of the experimental setup, data for coarse mesh number are only available for a relatively high Reynolds number. Similarly, data for fine mesh are only available for the low Reynolds number region. Therefore, it is possible that the phase shift between pressure and velocity seen within the coarse mesh is due to the relatively high velocity excited by the linear motor. This could be the reason for the time delay between pressure drop and velocity. It is likely that the phase shift between pressure and velocity is the feature of the inertial effect described by the Forchheimer model.

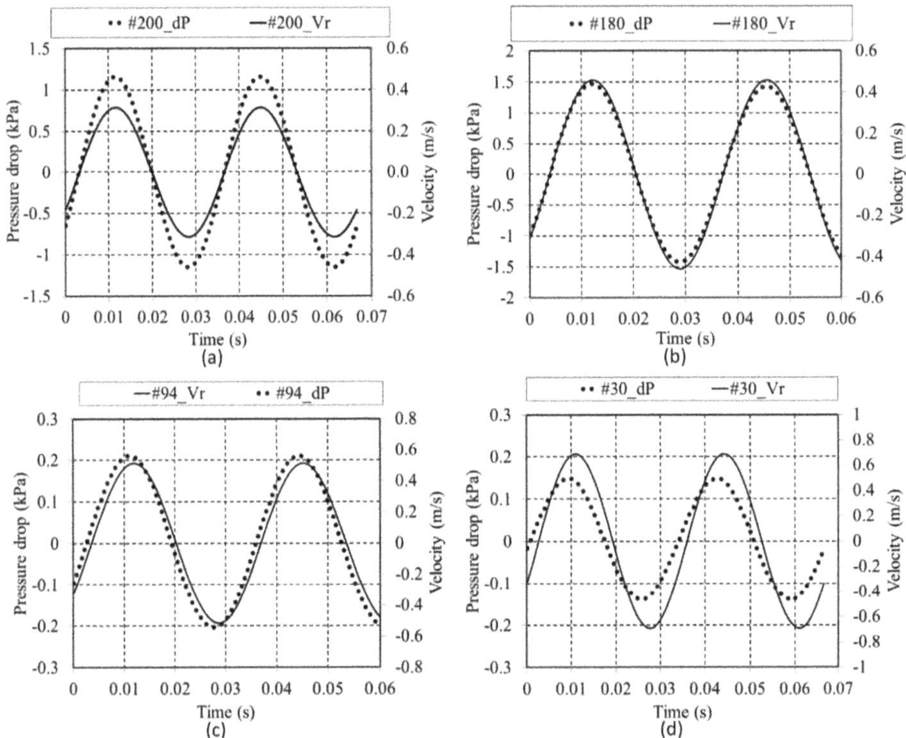

Figure 9. Pressure drop (dP) and velocity at regenerator (Vr) for: (**a**) #200; (**b**) #180; (**c**) #94; and (**d**) #30.

6. Conclusions

Modelling the full state of a regenerator may be very computationally demanding due to the very small size of pore and the tortuosity of the structure. Hence, in this study, an economical way of modelling a regenerator is developed. A two-dimensional model has been developed for the regenerator working in a travelling-wave setting. The structure is modelled as a porous medium. Modelling a porous medium requires a proper closure model to represent the momentum losses and thermal inertia of the flow due to the presence of the structure. In this study, the momentum losses are modelled with pressure drop data gained from the experiment conducted in a well-controlled travelling-wave condition.

The pressure drop of the porous tortuous structure of the regenerator has been modelled and validated by the results obtained from experiment carried out in an apparatus where a control could be achieved over the phase angle between velocity and pressure thus enabling a travelling-wave condition to be imposed. The values of computationally determined permeability coefficient, K, that represents the hydrodynamic loss of the regenerator investigated, are shown in Table 3. Friction factor correlation has been developed based on a porous medium model. A simplification offered by Darcy's law has been used for the low speed condition studied. The results from this correlation appear to match the results calculated through the correlation proposed by Swift and Ward [7]. This indicates that steady flow assumptions, used by them when developing their correlation, apply to the travelling-wave conditions investigated here.

This study suggests that the difference in friction correlation between oscillatory flow and steady flow, as reported in many previous studies, may be related to the "phase shifting" between pressure and velocity which are used for predicting the friction factor. For a travelling-wave system, when the pressure and velocity are in-phase, the pressure drop and velocity relate to each other in a similar way as in a steady flow. Results of this investigation show that the pore size of the mesh screen used in a travelling-wave system should not be too large. Otherwise, difficulties in sustaining the travelling-wave phase angle within the regenerator may arise. Should this happen the friction correlation may need to be revised to account for the effect of the "phase shifting" phenomenon.

Unfortunately, the results obtained from the experiment are somewhat limited because of the design of the apparatus. Flow within samples made of fine mesh screen was limited to low Reynolds number, while the flow within samples made of coarse mesh screen was at a relatively high Reynolds number (otherwise a sufficient pressure drop could not be recorded). Phase shift was observed between pressure and velocity within the regenerator made of coarse mesh screen (#94 and #30). The phase shifting phenomenon may be related to the inertial effect described by the Forchheimer model. This may be the reason for the discrepancy between numerical and experimental results at Reynolds number higher than 60.

The central focus of this study is pressure drop. However, thermal condition of the regenerator is also an important criterion to consider when selecting a regenerator. This can be done for example on the basis of Equation (24) in reference [3], while similar insights are also available in references [7,24]. A regenerator with a coarse mesh screen may be favourable in that it provides a smaller pressure drop. Nevertheless, the thermal contact may not be as good as for the fine mesh. Further investigation on this subject is strongly suggested to cover the whole range of the thermoacoustic phenomena, including the heat transfer within the regenerator.

Acknowledgments: The first author would like to acknowledge the sponsorship of the Ministry of Education, Malaysia and Universiti Teknikal Malaysia Melaka (FRGS/1/2015/TK03/FKM/03/F00274). The second author would like to acknowledge the sponsorship through the EPSRC Advanced Research Fellowship, grants GR/T04502 and GR/T04519, and Royal Society Industry Fellowship (2012-2015), grant number IF110094. Both would like to acknowledge Dr Yu for introducing the first author to the art of experimental work.

Author Contributions: Fatimah A.Z. Mohd Saat performed the experiments and numerical investigations, analysed the data and drafted the paper; Artur J. Jaworski planned and supervised the research works, outlined the paper and subsequently worked with first author on consecutive "iterations" of the paper till the final manuscript.

Conflicts of Interest: The authors declare no conflict of interest. The funding sponsors had no role in the design of the study; in the collection, analyses, or interpretation of data; in the writing of the manuscript, and in the decision to publish the results.

References

1. Rayleigh. The explanation of certain acoustical phenomena. *Nature* **1878**, 319–321.
2. Backhaus, S.; Swift, G.W. A thermoacoustic-Stirling heat engine: Detailed study. *J. Acoust. Soc. Am.* **2000**, *107*, 3148–3166. [CrossRef] [PubMed]
3. Yu, Z.; Jaworski, A.J. Impact of acoustic impedance and channel dimensions on the power output capacity of the regenerators in travelling-wave thermoacoustic engines. *Energy Convers. Manag.* **2010**, *51*, 350–359. [CrossRef]
4. Bejan, A.; Dincer, I.; Lorente, S.; Miguel, A.F.; Reis, A.H. *Porous and Complex Flow Structures in Modern Technologies*; Springer: New York, NY, USA, 2004; p. 9.
5. Choi, S.; Nam, K.; Jeong, S. Investigation on the pressure drop characteristic of cryoocooler regenerators under oscillating flow and pulsating pressure conditions. *Cryogenics* **2004**, *44*, 203–210. [CrossRef]
6. Gedeon, D.; Wood, J.G. *Oscillating-Flow Regenerator Test Rig: Hardware and Theory with Derived Correlations for Screens and Felt*; NASA Contractor Report 198442; NASA Lewis Research Center: Cleveland, OH, USA, 1996.
7. Swift, G.W.; Ward, W.C. Simple Harmonic Analysis of Regenerators. *J. Thermophys. Heat Transf.* **1996**, *10*, 652–662. [CrossRef]
8. Kays, W.M.; London, A.L. *Compact Heat Exchangers*; McGraw-Hill: New York, NY, USA, 1964.
9. Ward, W.C.; Swift, G.W. Design environment for low-amplitude thermoacoustic engines. *J. Acoust. Soc. Am.* **1994**, *95*, 3671–3672. [CrossRef]
10. Ju, Y.; Jiang, Y.; Zhuo, Y. Experimental study of the oscillating flow characteristics for a regenerator in a pulse tube cryocooler. *Cryogenics* **1998**, *38*, 649–656. [CrossRef]

11. Nam, K.; Jeong, S. Novel flow analysis of regenerator under oscillating flow with pulsating pressure. *Cryogenics* **2005**, *45*, 368–379. [CrossRef]
12. Boroujerdi, A.A.; Esmaeili, M. Characterization of the frictional losses and heat transfer of oscillatory viscous flow through wire mesh regenerators. *Alex. Eng. J.* **2015**, *54*, 787–794. [CrossRef]
13. Liu, S.; Chen, X.; Zhang, A.; Wu, Y.; Zhang, H. Investigation on phase shifter of a 10 W/70 K inertance pulse tube refrigerator. *Int. J. Refrig.* **2017**, *74*, 448–455. [CrossRef]
14. Hsu, C. Dynamic Modeling of Convective Heat Transfer in Porous Media. In *Handbook of Porous Media*, 2nd ed.; Vafai, K., Ed.; Taylor and Francis Group: Boca Raton, FL, USA, 2005; pp. 39–80.
15. Hsu, C.; Fu, H.; Cheng, P. On Pressure-Velocity Correlation of Steady and Oscillating Flows in Regenerators Made of Wire Screens. *J. Fluid Eng.* **1999**, *121*, 52–56. [CrossRef]
16. Hsu, C.T.; Cheng, P. Thermal dispersion in a porous medium. *Int. J. Heat Mass Transf.* **1990**, *33*, 1587–1597. [CrossRef]
17. Zhao, T.S.; Cheng, P. Oscillatory pressure drops through a woven-screen packed column subjected to a cyclic flow. *Cryogenics* **1996**, *36*, 333–341. [CrossRef]
18. Cha, J.S.; Ghiaasiaan, S.M.; Kirkconnell, C.S. Oscillatory flow in microporous media applied in pulse-tube and Stirling-cycle cryocooler regenerators. *Exp. Therm. Fluid Sci.* **2008**, *32*, 1264–1278. [CrossRef]
19. Clearman, W.M.; Cha, J.S.; Ghiaasiaan, S.M.; Kirkconnell, C.S. Anisotropy steady-flow hydrodynamic parameters of microporous media applied to pulse tube and Stirling cryocooler regenerators. *Cryogenics* **2008**, *48*, 112–121. [CrossRef]
20. Landrum, E.C.; Conrad, T.J.; Ghiaasiaan, S.M.; Kirkconnell, C.S. Hydrodynamic parameters of mesh fillers relevant to miniature regenerative cryocoolers. *Cryogenics* **2010**, *50*, 373–380. [CrossRef]
21. Tao, Y.B.; Liu, Y.W.; Gao, F.; Chen, X.Y.; He, Y.L. Numerical analysis on pressure drop and heat transfer performance of mesh regenerators used in cryocoolers. *Cryogenics* **2009**, *49*, 497–503. [CrossRef]
22. Gardner, D.L.; Swift, G.W. Use of inertance in orifice pulse tube refrigerators. *Cryogenics* **1997**, *37*, 117–121. [CrossRef]
23. Song, B.H.; Bolton, J.S. A transfer-matrix approach for estimating the characteristic impedance and wave number of limp and rigid porous materials. *J. Acoust. Soc. Am.* **2000**, *167*, 1131–1152. [CrossRef]
24. Ueda, Y.; Kato, T.; Kato, C. Experimental evaluation of the acoustic properties of stacked-screen regenerators. *J. Acoust. Soc. Am.* **2009**, *125*, 780–786. [CrossRef] [PubMed]
25. Swift, G.W. *Thermoacoustics: A Unifying Perspective for Some Engines and Refrigerators*; Acoustical Society of America Publications: Sewickley, PA, USA, 2002.
26. ANSYS FLUENT 13.0. *User Manual*, ANSYS Inc.: Canonsburg, PA, USA, 2010.

© 2017 by the authors. Licensee MDPI, Basel, Switzerland. This article is an open access article distributed under the terms and conditions of the Creative Commons Attribution (CC BY) license (http://creativecommons.org/licenses/by/4.0/).

Article

Excitation of Surface Waves Due to Thermocapillary Effects on a Stably Stratified Fluid Layer

William B. Zimmerman [1,†] and Julia M. Rees [2,*,†]

1 Department of Chemical and Biological Engineering, Mappin Street, University of Sheffield, Sheffield S1 3JD, UK; w.zimmerman@sheffield.ac.uk
2 School of Mathematics and Statistics, Hounsfield Road, University of Sheffield, Sheffield S3 7RH, UK
* Correspondence: j.rees@sheffield.ac.uk; Tel.: +44-114-222-3782
† These authors contributed equally to this work.

Academic Editor: Artur J. Jaworski
Received: 23 February 2017; Accepted: 10 April 2017; Published: 13 April 2017

Abstract: In chemical engineering applications, the operation of condensers and evaporators can be made more efficient by exploiting the transport properties of interfacial waves excited on the interface between a hot vapor overlying a colder liquid. Linear theory for the onset of instabilities due to heating a thin layer from above is computed for the Marangoni–Bénard problem. Symbolic computation in the long wave asymptotic limit shows three stationary, non-growing modes. Intersection of two decaying branches occurs at a crossover long wavelength; two other modes co-exist at the crossover point—propagating modes on nascent, shorter wavelength branches. The dispersion relation is then mapped numerically by Newton continuation methods. A neutral stability method is used to map the space of critical stability for a physically meaningful range of capillary, Prandtl, and Galileo numbers. The existence of a cut-off wavenumber for the long wave instability was verified. It was found that the effect of applying a no-slip lower boundary condition was to render all long waves stationary. This has the implication that any propagating modes, if they exist, must occur at finite wavelengths. The computation of 8000 different parameter sets shows that the group velocity always lies within $\frac{1}{2}$ to $\frac{2}{3}$ of the longwave phase velocity.

Keywords: thermocapillary; Marangoni number; stability analysis

---

1. Introduction

Marangoni–Bénard convection still generates much interest in fluid and nonlinear dynamics due to its complexity. When the fluid layer is heated from below, convective instabilities can be driven by surface or buoyancy forces [1]. The role of surface-tension gradients in inducing convective instability through Marangoni stresses at the air–liquid interface in a thin layer initially at rest, heated from below, was characterized in the seminal works of Sternling and Scriven [2] and Smith [3] and the role of surface deformation and surface tension gradients in the onset of patterned convection and oscillatory instability is reviewed in [4].

When the layer is heated from above, however, only overstability can be excited at sufficiently high Marangoni or Rayleigh (buoyancy) numbers. The most common chemical engineering context for a cold liquid heated from above is a condenser, which comprises a hot vapor that condenses over a colder liquid chilled from the solid support below. There are many different configurations for condensers, but all of them would condense faster if interfacial waves are excited on the interface. Thus, a stability theory that shows under what conditions self excitation occurs could better inform the design and operation of condensers. The scenario also describes an evaporator that is operated by contacting hot gas over the cold liquid, as long as there is no fluid motion imposed. Of course, imposing fluid motion adds additional complication but has been found to accelerate performance in novel contactors (evaporators, condensers and distillers) [5].

Classical experiments by Linde and coworkers [6] demonstrate a series of wave instabilities excited at high Marangoni numbers, although they could not discern whether the excited waves were surface or internal waves. It was later posited that the waves were surface manifestations of soliton solutions to a dissipative variation of the Korteweg–de Vries equation. Experiments have shown that solitary waves, excited and sustained by Marangoni stresses, undergo interactions and collisions that return the waves to the pre-collision celerities and shapes, experiencing at most a phase shift, but with either sense possible [7]. Nepomnyashchy and Velarde [8] demonstrated via multiple scale perturbation methods the definitive derivation of the dissipative nonlinear evolution equation (termed the KdV–KSV equation) in a sufficiently thin layer that buoyancy effects are neglected. Their study assumed a stress-free lower boundary. The KdV–KSV theory predicts a critical Marangoni number $M_{crit} = 12$, irrespective of capillary, Prandtl and gravity numbers. The interpretation of this result is simply that surface-tension gradients, if sufficiently strong, overcome viscous dissipation in the surface layer to start the fluid oscillations. These oscillations must then combine two modes—the gravity waves modified by Marangoni stresses that are normal to the surface (termed transverse waves) and elastic-like waves that are tangential to the surface (termed longitudinal waves). A study of solutocapillary Marangoni-induced interfacial waves is given in [9].

The purpose of this paper is to test the regime of validity of the KdV–KSV theory on two points—(i) the assumption of the multiple scale theory that there is a long wave cut-off for a wave-packet of unstable waves just above critical stability; (ii) the affect of the no-slip lower boundary on the coefficients of the linear terms of the KdV–KSV equation—by computing the full linear stability theory (LST) of the Bénard–Marangoni problem when heated from above. Since, in the basic state, the fluid is at rest and only the static pressure among the field variables has a vertical dependency, it is possible to formulate the analytic solution to the linearized equations and subsequently the secular equation for excitation of non-trivial solutions for the streamfunction, temperature, and surface displacement. The analysis presented is a precursor to the weakly nonlinear theory for the evolution of capillary gravity waves given in [10]. The model equations are presented in Section 2. In Section 3, the linear stability of both the stress-free model and the no-slip models are formulated. In each case, the long wave asymptotic forms are computed for branches of the dispersion relation found. Results are discussed in Section 4 and the conclusions are drawn in Section 5.

2. Flow Specification

2.1. Scaling

Consider a layer of viscous fluid with dynamic viscosity μ, density ρ, and thermal diffusivity κ, which is heated from above. The background no flow and no deformation state is perturbed, giving rise to gravity, capillary, and Marangoni forces. The relevant physical parameters are summarized below:

- Surface tension $\sigma = \sigma_0 + \gamma (T - T_0)$, where σ_0 is the surface tension at the reference temperature T_0.
- Surface height $z = d + h(x, y, t)$, where d is the nominal height of the surface in the absence of deformation, and x and y are horizontal coordinates. t is time.
- Stationary temperature profile $T_s = T_0 + \beta (z - d)$.
- Hydrostatic pressure $p_s = P_0 - \rho g (z - d)$.
  As the focus here is on surface forces, the buoyancy in the bulk will be neglected by assuming density constant.

We take diffusive scales as follows: $t^* = \frac{\kappa t}{d^2}$, $x_i^* = \frac{x_i}{d}$, $v_i^* = \frac{d v_i}{\kappa}$, $T^* = 1 + \frac{T - T_0}{\beta d}$, $\sigma^* = \frac{\sigma}{\sigma_0}$, $p^* = \frac{d^2 p}{\mu \kappa}$, $\sigma_{ij}^* = \frac{d^2 \sigma_{ij}}{\mu \kappa}$, where σ_{ij} is the stress tensor. The asterisks refer to dimensionless variables. Henceforth, the asterisks will be dropped and all calculations are given in dimensionless variables.

Dimensional analysis yields four dimensionless groups:

- Prandtl number $\mathrm{Pr} = \frac{\nu}{\kappa}$,
- Galileo number $G = \frac{gd^3}{\nu\kappa}$,
- Capillary number $K = \frac{\mu\kappa}{\sigma_0 d}$,
- Marangoni number $M = -\frac{\gamma\beta d^2}{\mu\kappa}$.

As the layer is heated from above, β will be negative for simple fluids and the Marangoni number as defined above is an intrinsically positive quantity for the target situation.

2.2. Model Equations

The full governing equations are adapted from Davis and Homsy [11] by neglecting buoyancy. In the bulk, the velocity, temperature and pressure are constrained as:

$$\begin{aligned} \frac{1}{\mathrm{Pr}}\left(\frac{\partial v_i}{\partial t} + v_j v_{i,j}\right) &= \sigma_{ij}, \\ \frac{\partial T}{\partial t} + v_i T_{,i} &= \nabla^2 T, \\ v_{i,i} &= 0, \end{aligned} \tag{1}$$

where $\sigma_{ij} = -p\delta_{ij} + \epsilon_{ij}$; $\epsilon_{ij} = v_{i,j} + v_{j,i}$. The comma-subscript represents the index convention for partial differentiation with respect to x_i. The boundary conditions are a rigid lower planar surface held at constant temperature at $z = 0$:

$$\begin{aligned} v_i &= 0, \\ T &= 0. \end{aligned}$$

The upper free surface is open and deformable at $z = 1 + \eta\,(x, y, t)$:

$$\begin{aligned} \eta_t &= N v_i n_i, \\ \sigma_{ij} n_j &= \frac{\mathcal{K}(\eta)}{K}\sigma n_i - M T_{,k} t_k t_i, \\ T_{,i} n_i &= 1. \end{aligned} \tag{2}$$

The differential geometry of the surface $S(t) : z = 1 + \eta$ is given in terms of the element of arc length $N(\eta)$, curvature $\mathcal{K}(\eta)$, and the normal n_i and the tangent t_i vectors:

$$\begin{aligned} N(\eta) &= \left(1 + \eta_x^2\right)^{1/2}, \\ \mathcal{K}(\eta) &= \tfrac{\eta_{xx}}{N}, \\ \mathbf{n} &= (-\eta_x, 1)N^{-1}, \\ \mathbf{t} &= (1, \eta_x)N^{-1}. \end{aligned} \tag{3}$$

The base state is motionless with only hydrostatic pressure:

$$\begin{aligned} \hat{T} &= z, \\ \hat{\sigma}_{ij} &= G\,(z-1)\,\delta_{ij}, \\ \hat{\eta} &= 0. \end{aligned} \tag{4}$$

3. Linearized System

The linearized equations about the motionless base state (4) for a two-dimensional disturbance can be written in terms of a streamfunction ψ, defined in the usual way for incompressible flow. In dimensionless form, the streamfunction–vorticity equation is given by:

$$\frac{1}{\mathrm{Pr}}\left(\psi_{zzt}+\psi_{xxt}\right)=\psi_{xxxx}+2\psi_{xxzz}+\psi_{zzzz}. \tag{5}$$

Subscripting by coordinates refers to partial differentiation by the respective coordinate. The pressure, which is needed for the normal stress boundary condition, can be found from

$$p_x=\psi_{xxz}+\psi_{zzz}-\frac{1}{\mathrm{Pr}}\psi_{zt}. \tag{6}$$

The mathematical analysis leading to Equations (5) and (6) is outlined in Davis and Homsy [11]. The linear stability analysis is a standard mathematical approach having linearized the equations about the base state. Heat transport couples temperature convection and diffusion. The linearized version is

$$T_t=T_{xx}+T_{zz}+\psi_x. \tag{7}$$

We apply seven boundary conditions at the upper and lower surfaces.

At $z=0$, material surface (no penetration):

$$\psi=0; \tag{8}$$

no slip:

$$\psi_z=0; \tag{9}$$

fixed temperature:

$$T=0. \tag{10}$$

At $z=1$, normal stress:

$$p=-2\psi_{xz}+Gh-K^{-1}h_{xx}; \tag{11}$$

tangential stress:

$$0=\psi_{zz}-\psi_{xx}+M\left(T_x+h_x\right); \tag{12}$$

material surface (kinematic condition):

$$h_t=-\psi_x; \tag{13}$$

no heat flux:

$$T_z=0. \tag{14}$$

We presume separation of variables with a factor that is a wave of real wavenumber k and (complex) phase velocity c by normal mode expansion of the field variables:

$$\begin{aligned} p &= P(z)\exp\left(ik\left(x-ct\right)\right), \\ \psi &= \Psi(z)\exp\left(ik\left(x-ct\right)\right), \\ T &= \Theta(z)\exp\left(ik\left(x-ct\right)\right), \\ h &= H\exp\left(ik\left(x-ct\right)\right). \end{aligned} \tag{15}$$

The wave system is thus reduced to a two-point boundary value problem of ordinary differential equations (ODEs) in $P(z), \Psi(z), \Theta(z)$ and H, with algebraic constraints: Bulk equations,

$$\Psi_{zzzz} + \left(\frac{ikc}{\mathrm{Pr}} - 2k^2\right)\Psi_{zz} + \left(k^4 - \frac{ik^3c}{\mathrm{Pr}}\right)\Psi = 0, \tag{16}$$

$$P = ik\Psi_z - \frac{i}{k}\Psi_{zzz} + \frac{c}{\mathrm{Pr}}\Psi_z, \tag{17}$$

$$\Theta_{zz} + \left(ikc - k^2\right)\Theta = -ik\Psi. \tag{18}$$

Boundary conditions, at $z = 0$,

$$\begin{aligned} \Psi &= 0, \\ \Psi_z &= 0, \\ \Theta &= 0. \end{aligned} \tag{19}$$

At $z = 1$,

$$\begin{aligned} P &= -2ik\Psi_z + \left(G + k^2/K\right)H, \\ 0 &= \Psi_{zz} + k^2\Psi + Mik\left(\Theta + H\right), \\ \Theta_z &= 0, \\ cH &= \Psi. \end{aligned} \tag{20}$$

The algebraic manipulation package Mathematica 5.2 (Wolfram, Champaign, IL, USA) was used to manipulate the governing equations for this system. As the bulk equations are linear ODEs with constant coefficients, the general solutions can be represented by linear combinations of complex exponentials given by the characteristic roots, $\pm k \pm \lambda = \sqrt{k^2 - \frac{ikc}{\mathrm{Pr}}}$ and $\pm\Lambda = \sqrt{k^2 - ikc}$, yielding

$$\begin{aligned} \Psi &= A\exp(kz) + B\exp(-kz) + C\exp(\lambda z) + D\exp(-\lambda z), \\ \Theta &= \frac{-ik}{k^2 - \Lambda^2}\left(A\exp(kz) + B\exp(-kz)\right) + \frac{-ik}{\lambda^2 - \Lambda^2}\left(C\exp(\lambda z) + D\exp(-\lambda z)\right), \\ &+ F\exp(\Lambda z) + J\exp(-\Lambda z). \end{aligned} \tag{21}$$

The seven boundary conditions (19) and (20) reduce to a matrix equation in the seven unknowns $\alpha = [A, B, C, D, F, J, H]^T$, $\mathbf{Z}\alpha = 0$, where the matrix $\mathbf{Z} = \mathbf{Z_{stick}}$ is given by

$$\mathbf{Z_{stick}} = \begin{bmatrix} 1 & 1 & 1 & 1 & 0 & 0 & 0 \\ \sqrt{k} & -\sqrt{k} & \frac{\lambda}{\sqrt{k}} & -\frac{\lambda}{\sqrt{k}} & 0 & 0 & 0 \\ -1 & -1 & \frac{\mathrm{Pr}}{1-\mathrm{Pr}} & \frac{\mathrm{Pr}}{1-\mathrm{Pr}} & c & c & 0 \\ \frac{ke^k(c+2ik\mathrm{Pr})}{\mathrm{Pr}} & -\frac{ke^{-k}(c+2ik\mathrm{Pr})}{\mathrm{Pr}} & 2ike^{\lambda}\lambda & -2ike^{-\lambda}\lambda & 0 & 0 & -G - \frac{k^2}{K} \\ e^k\left(2k - \frac{iM}{c}\right) & e^{-k}\left(2k - \frac{iM}{c}\right) & e^{\lambda}Q & e^{-\lambda}Q & iMe^{\Lambda} & iMe^{-\Lambda} & iM \\ -\frac{\sqrt{k}e^k}{c} & \frac{\sqrt{k}e^{-k}}{c} & -\frac{e^{\lambda}\lambda}{c\sqrt{k}(1-\frac{1}{\mathrm{Pr}})} & \frac{e^{-\lambda}\lambda}{c\sqrt{k}(1-\frac{1}{\mathrm{Pr}})} & \frac{e^{\Lambda}\Lambda}{\sqrt{k}} & -\frac{e^{-\Lambda}\Lambda}{\sqrt{k}} & 0 \\ e^k & e^{-k} & e^{\lambda} & e^{-\lambda} & 0 & 0 & -c \end{bmatrix}, \tag{22}$$

where

$$Q = \left(2k - \frac{ic}{\mathrm{Pr}} - \frac{iM}{c(1 - \frac{1}{\mathrm{Pr}})}\right).$$

Alternatively, with slip boundary conditions, where boundary condition (9) is replaced by the stress-free condition on the lower surface $\psi_{zz} = 0$, we have the matrix $\mathbf{Z} = \mathbf{Z_{slip}}$:

$$\mathbf{Z}_{\mathbf{slip}} = \begin{bmatrix} 1 & 1 & 1 & 1 & 0 & 0 & 0 \\ k & k & \lambda^2/k & \lambda^2/k & 0 & 0 & 0 \\ -1 & -1 & \frac{\mathrm{Pr}}{1-\mathrm{Pr}} & \frac{\mathrm{Pr}}{1-\mathrm{Pr}} & c & c & 0 \\ \frac{ke^k(c+2ik\,\mathrm{Pr})}{\mathrm{Pr}} & -\frac{ke^{-k}(c+2ik\,\mathrm{Pr})}{\mathrm{Pr}} & 2ike^{\lambda}\lambda & -2ike^{-\lambda}\lambda & 0 & 0 & -G-\frac{k^2}{K} \\ e^k\left(2k-\frac{iM}{c}\right) & e^{-k}\left(2k-\frac{iM}{c}\right) & e^{\lambda}Q & e^{-\lambda}Q & iMe^{\Lambda} & iMe^{-\Lambda} & iM \\ -\frac{\sqrt{k}e^k}{c} & \frac{\sqrt{k}e^{-k}}{c} & -\frac{e^{\lambda}\lambda}{c\sqrt{k}(1-\frac{1}{\mathrm{Pr}})} & \frac{e^{-\lambda}\lambda}{c\sqrt{k}(1-\frac{1}{\mathrm{Pr}})} & \frac{e^{\Lambda}\Lambda}{\sqrt{k}} & -\frac{e^{-\Lambda}\Lambda}{\sqrt{k}} & 0 \\ e^k & e^{-k} & e^{\lambda} & e^{-\lambda} & 0 & 0 & -c \end{bmatrix}. \quad (23)$$

3.1. Slip Boundary Condition

Either we have $\alpha = 0$, in which case no wave solution exists, or there is a non-trivial solution with a dispersion relation for the phase velocity as a function of wavenumber $c(k)$ implied by the singularity of the matrix, $\Delta = det(\mathbf{Z}) = 0$. Expanding the determinant Δ algebraically using the permutation rule gives rise to 720 complex exponential terms upon eliminating H through application of the kinematic condition on the upper surface. Thus, the determination of the dispersion relation through solving $\Delta = 0$ symbolically is not a trivial undertaking. In contrast, the computation is readily tractable numerically, apart from a particular difficulty in finding the zeros of Δ. Numerical computations with fixed precision arithmetic are unlikely to return exactly zero. Therefore, a small number below a given threshold is typically taken to be zero. When the determinant of a matrix approaches zero, its condition number becomes large and thus the matrix becomes ill-conditioned, meaning that computations could involve high levels of numerical error. Consequently, the threshold level required to identify zeros of Δ can be difficult to determine a priori. This motivates our desire to seek closed form symbolic approximations to the dispersion relation.

Such approximations usually start with the long wave limiting cases. Presuming that c is an analytic function of k allows us to write the Taylor's series expansion of $\Delta(c,k)$:

$$\Delta(c,k) = \sum \Delta_n(c,0)\,k^n. \quad (24)$$

Truncating the series at $n = N$ yields an approximate dispersion relation implicitly by requiring the approximate determinant to vanish. We note that, with the matrix $\mathbf{Z}_{\mathbf{slip}}$ as written, the first non-trivial contribution comes at $n = 3/2$, with

$$\Delta_{3/2} = -8iPr^{7/2}c^4\left(c^2 - (G+M)\,Pr\right) \quad (25)$$

having a quadruple root at $c = 0$ and real roots

$$c = \pm\sqrt{(G+M)\,Pr}. \quad (26)$$

These are the long wave asymptotic limits that demonstrate that the Marangoni effect is additive to the gravity wave effect in contributing to the phase velocity. The capillary effect, however, is classically weaker by $O(k^2)$ [12].

To isolate these modes, it is convenient to translate the phase velocity to the frame of reference of the leftward (or rightward) moving wave and consider $\Delta\left(\delta c + \sqrt{(G+M)\,\mathrm{Pr}}, k\right) = 0$. Since, by construction, Δ is analytic in both k and c, it is useful to develop the k-power series of the c-truncation of Δ, i.e.,

$$\Delta\left(\delta c + \sqrt{(G+M)\,\mathrm{Pr}}, k\right) \approx \Delta\left(\sqrt{(G+M)\,\mathrm{Pr}}, k\right) + \delta c\Delta^{(1)}\left(\sqrt{(G+M)\,\mathrm{Pr}}, k\right) = 0. \quad (27)$$

Developing Equation (27) to $O(k^4)$ yields

$$\delta c = q_1 k + q_2 k^2 + q_3 k^3, \quad (28)$$

with

$$
\begin{aligned}
q_1 &= -\frac{i}{6}(M-12)\,\mathrm{Pr},\\
q_2 &= -\frac{\sqrt{\mathrm{Pr}}\left(-45+3GK\left(5+2M\,\mathrm{Pr}\right)+K\left(30M+M^2+1440\,\mathrm{Pr}-240M\,\mathrm{Pr}+16M^2\,\mathrm{Pr}\right)\right)}{90K\sqrt{G+M}},\\
q_3 &= \frac{i}{7560K\,(G+M)}(4410\,(12-M)\,\mathrm{Pr}+204G^2KM\mathrm{Pr}^2\\
&+ GK\left(-2520\,\mathrm{Pr}\,(11+6\,\mathrm{Pr})-42M\left(-4-85\,\mathrm{Pr}+156\mathrm{Pr}^2\right)+M^2\left(8+29\,\mathrm{Pr}+1059\mathrm{Pr}^2\right)\right)\\
&+ K(-15120\,\mathrm{Pr}\,(1+65\,\mathrm{Pr})+2520M\,\mathrm{Pr}\,(-17+91\,\mathrm{Pr})\\
&- 21\,M^2\left(-8-179\,\mathrm{Pr}+1277\,\mathrm{Pr}^2\right)+M^3\left(8+127\,\mathrm{Pr}+1415\,\mathrm{Pr}^2\right))),
\end{aligned}
$$

which is readily interpreted as a long wave instability occurring for a Marangoni number greater than the critical value $M_{crit} = 12$ for long waves $k << 1$. As $k = 0$ is a critical wavenumber with neutral stability, it follows that for the problem to be well posed in the sense of Joseph and Saut [13], $Im\{c(k)\} < 0$ as $k \to \infty$. Thus, there must be a cut-off wavenumber for the long wave instability, where $Im\{c(k_{cutoff})\} = 0$ for $k > 0$ and higher wavenumbers decay, presumably due to the viscous and heat dissipation modes. Conveniently, the exponential growth rate with slightly supercritical M grows quadratically in k^2 (Marangoni pumping) and decays as k^4, since the growth constant is $-kIm\{c\}$. As the odd order contributions in k are purely imaginary, it is possible that the long wave instability is limited to a wave packet, with higher wavenumbers than k_{cutoff} being damped. Figure 1 demonstrates this occurrence for slightly supercritical $M = 12.1$. It can be seen that the growth rate is positive for wavenumbers less than approximately 0.007, but that for larger wavenumbers, the growth rate is negative, which is indicative of decaying modes.

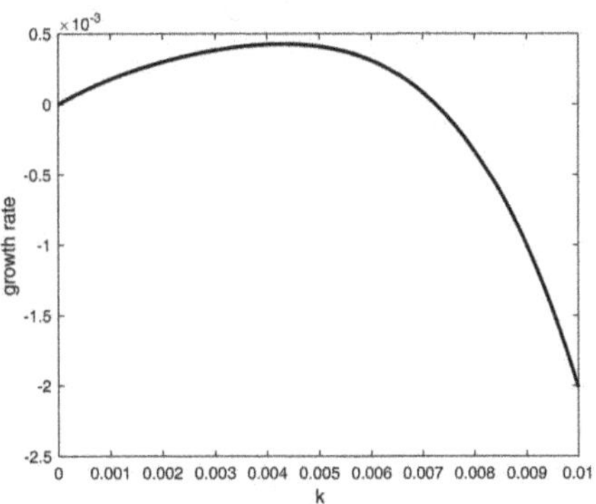

Figure 1. Growth rate $-Im\{c\}$ vs. wavenumber k for $M = 12.1$, $\mathrm{Pr} = 10$, $G = 100$, $K = 0.00001$. The plot clearly illustrates an exponential growth of a long wave packet, subtended by $k = 0$ and k_{cutoff}.

An analytic form for the cutoff wavenumber can be computed from Equation (28), which also shows the supercritical nature of the long wave packet of instability. Ref. [8] assumed the occurrence of just this form of dispersion relation as the basis for their multiple scale perturbation theory. They assumed $k_{cutoff} << 1$ as the formal perturbation parameter. Given that $k_{cutoff} = 0.00724$ for the conditions in Figure 1, their intuition has been proved correct here.

3.2. No-Slip Boundary Conditions

Introducing the no-slip lower boundary makes a major structural change to the physics of the problem. It is well known that there are three fluid dynamical dissipative mechanisms that lead to the attenuation of surface waves. *Bottom friction* is the dominant mechanism wherever the layer depth is substantially less than the wavelength, so that the wave induces large horizontal motions near the bottom (see [12], Figure 55 for an illustration of this point).

Deep fluid waves, however, do not induce movement near the bottom, and thus no frictional dissipation. *Internal dissipation* by viscous stresses acting throughout the wave cause attentuation.

Surface dissipation is associated with departures of the surface from its equilibrium value, described, for example, by Lucassen (1968) in the case of doping of the surface with a monolayer of surfactant leading to an elastic dissipative mechanism.

In addition to these mechanisms, the configuration under study has internal dissipation from the thermal conduction and possible surface dissipation from the Marangoni effect, which can play the role of a Lucassen-like tangential surface stress.

Lighthill [12] estimated the proportional energy loss per period, due solely to the bottom friction, as

$$\left[\frac{2\pi}{d}\sqrt{\frac{\nu}{2\Omega}}\right]\left\{\frac{2kd}{sinh(2kd)}\right\}, \tag{29}$$

where $\Omega = c/k$ is the wave frequency. The square bracket factor is the ratio of the thickness of the bottom viscous boundary layer induced by the wave to the depth. The curly bracket factor corrects for finite wavenumber—it is unity for infinitely long waves and zero for infinitely short ripples.

A complementary analysis for internal dissipation leads to the opposite preferences. Infinitely long waves are unaffected by internal dissipation; infinitely short waves are massively damped by internal dissipation. No doubt that consideration of internal dissipation effects only influenced the search by [8] for long wave excitation of surface solitary waves.

A priori, this estimate would suggest that excitation of solitary disturbances cannot occur for either long waves or short waves due to the dominance of the attenuation by bottom friction and internal dissipative mechanisms, respectively. Candidate wavenumbers for solitary wave excitation should be intermediate wavenumbers where the Marangoni effect induces sufficient disturbance energy to overcome internal dissipation and bottom friction.

To investigate this hypothesis, the modal structure of the no-slip problem should be clarified. The principal reason is that the major branches can be described with analytic approximations, leading to better understanding of the structure, and highlighting changes that are possible at intermediate wavenumbers.

3.3. No-Slip Boundary Condition: Modal Structure

It is now convenient to express separation of variables for the no-slip problem for each normal mode in terms of exponential factors appropriate for growing standing waves (ω real):

$$\begin{aligned} \psi(x,z,t) &= \Psi(z)\exp(ikx)\exp(\omega t), \\ T(x,z,t) &= \Theta(z)\exp(ikx)\exp(\omega t), \\ p(x,z,t) &= P(z)\exp(ikx)\exp(\omega t). \end{aligned} \tag{30}$$

The general solution can be found as:

$$\begin{aligned} \Psi &= \sum_{j=1}^{4}\alpha_j\exp(\lambda_j z), \\ T &= -ik\sum_{j=1}^{4}\frac{\alpha_j}{\lambda_j^2-\lambda_5^2}\exp(\lambda_j z)+\sum_{j=5}^{6}\alpha_j\exp(\lambda_j z), \\ P &= i\left(k+\frac{\omega}{\mathrm{Pr}\,k}\right)\frac{d\Psi}{dz}-\frac{i}{k}\frac{d^3\Psi}{dz^3}, \end{aligned} \tag{31}$$

where $\lambda_1 = k$, $\lambda_2 = -k$, $\lambda_3 = \left(k^2 + \frac{\omega}{\mathrm{Pr}}\right)^{1/2}$, $\lambda_4 = -\lambda_3$, $\lambda_3 = \left(k^2 + \omega\right)^{1/2}$ and $\lambda_6 = -\lambda_5$ are the characteristic values of the general solution.

The modal structure can be examined by considering the long wave limit

$$\lim_{k\to 0} \Delta\left(\omega, k; \mathrm{Pr}, M, K, G\right) = 0 \tag{32}$$

implicitly defines $\omega\left(k \to 0; \mathrm{Pr}, M, K, G\right)$. This limit can be determined symbolically from:

$$\lim_{k\to 0} \Delta\left(\omega, k; \mathrm{Pr}, M, K, G\right) = \frac{-2\,e^{-\sqrt{\omega}-\sqrt{\frac{\omega}{\mathrm{Pr}}}}\left(1+e^{2\sqrt{\omega}}\right)\left(1+e^{2\sqrt{\frac{\omega}{\mathrm{Pr}}}}\right)\omega^{\frac{13}{2}}\sqrt{\frac{\omega}{\mathrm{Pr}}}}{\mathrm{Pr}^2}. \tag{33}$$

Three finite roots exist: $\{\{\omega \to 0\}, \{\omega \to \frac{-\pi^2}{4}\}, \{\omega \to \frac{-(\pi^2\,\mathrm{Pr})}{4}\}\}$. We find that the neutrally stable root $\omega \to 0$ has multiplicity of at least seven. The root $\omega \to \frac{-\pi^2}{4}$ is associated with exponential decay. Since the scaling for time is diffusive, the exponential decay constant is a factor of the thermal time scale. Thus, this mode is termed the thermal mode. As the root $\omega \to \frac{-\pi^2\,\mathrm{Pr}}{4}$ decays according to the viscous time scale, we label this the viscous mode.

A key property of all of these modes is their stationarity. The long wave slip modes comprise a pair of propagating simple waves. The no-slip bottom boundary condition has the effect of rendering all long waves stationary. Thus, if any propagating modes exist, they must occur at finite wavelengths. This suggests investigating the k-dependence for long waves branching from the infinitely long wave modes identified above.

4. Results and Discussion

4.1. k-Space Continuation and Modal Character

Identifying the critical surface in $\{M, K, Pr, G\}$ parameter space for the co-existence of the pair of intermediate wavenumber propagating modes and the long wave stationary modes (long wave thermal mode and long wave neutral temperature mode) is essential in mapping the parameter space. Continuation methods based on Newton iteration must search for either purely real ω (long wave stationary modes) or genuinely complex ω (intermediate wavenumber propagating modes). Figure 2 shows this critical curve $k - M$ and $\omega - M$ for four mode co-existence at fixed parametric values $\{K = 0.0001, Pr = 1.001, G = 1\}$ found by numerically solving for $\Delta = 0$ and $\Delta_\omega = 0$ simultaneously by Newton's method. For $M \in \{500, 1500\}$, it can be seen that the wavenumber of the crossover point is a decreasing function, whilst the growth rate is an increasing function. The four modes that co-exist along this critical curve are the two long wave stationary modes and two intermediate modes propagating from opposing directions. Figure 3 shows an analogous $k - \mathrm{Pr}$ curve for the crossover point from stationary long waves to propagating steady intermediate waves. A Prandtl number of 1 is common for gases and a Prandtl number of 10 is common for liquids. The plots in Figure 3 show how the behavior of the fluids changes from a near gaseous state to a near liquid state.

Once the critical surface dividing the wavenumber space into the two different regions of wave character has been identified, it is a simple matter to use parameter space continuation methods in k to find the remainder of the dispersion relation. Figure 4 summarizes the dispersion relation by plotting maximum growth rate and the wavenumber at which it is achieved for a given Marangoni number M at the values $\{K = 0.0001, Pr = 1.001, G = 1\}$. There is jaggedness in the graph in Figure 4a as the wavenumber of maximum growth can switch modes. The normalized group velocity is also plotted as a function of the Marangoni number. It is clear that neutral stability is found by ramping up the Marangoni number to $M \sim 1100$ for the other parameters fixed at these values.

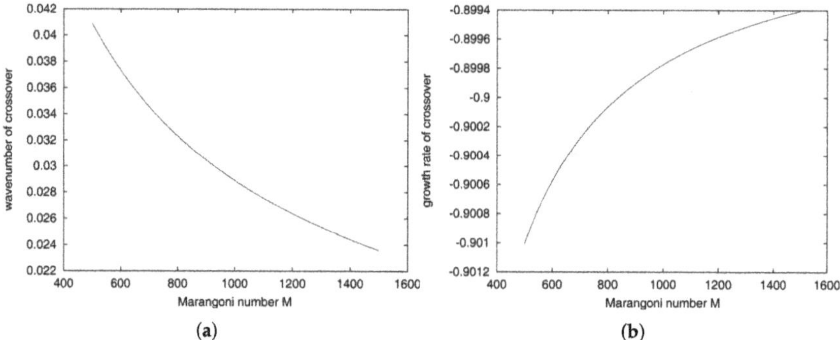

Figure 2. Magnitude of velocity field on $z = 0$ mm plane: (**a**) crossover wavenumber k; (**b**) growth rate ω; for which $\Delta = 0$ and $\Delta_\omega = 0$ simultaneously at the parametric values $\{K = 0.0001, Pr = 1.001, G = 1\}$ and $M \in \{500, 1500\}$. Along this critical curve, four modes co-exist—the two long wave stationary modes and two opposite-directed intermediate propagating modes.

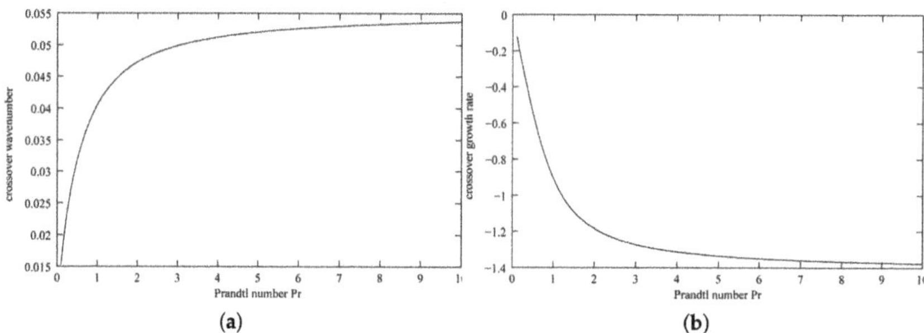

Figure 3. Magnitude of velocity field on $z = 0$ mm plane: (**a**) crossover wavenumber k; (**b**) growth rate ω; for which $\Delta = 0$ and $\Delta_\omega = 0$ simultaneously at the parametric values $\{K = 0.0001, M = 500, G = 1\}$ and $\mathrm{Pr} \in \{0.1, 10\}$. Along this critical curve, four modes co-exist—the two long wave stationary modes and two opposite-directed intermediate propagating modes.

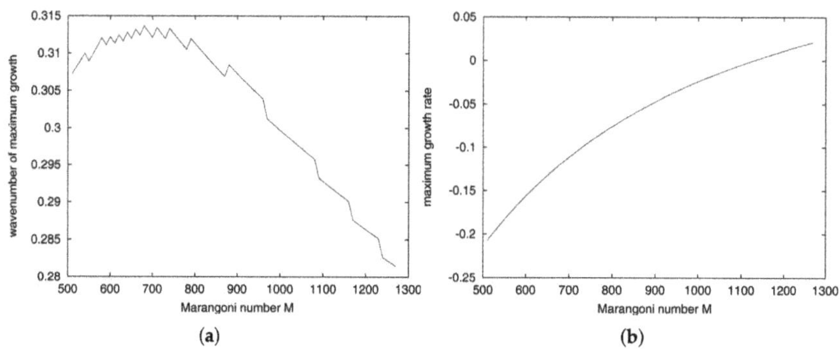

Figure 4. *Cont.*

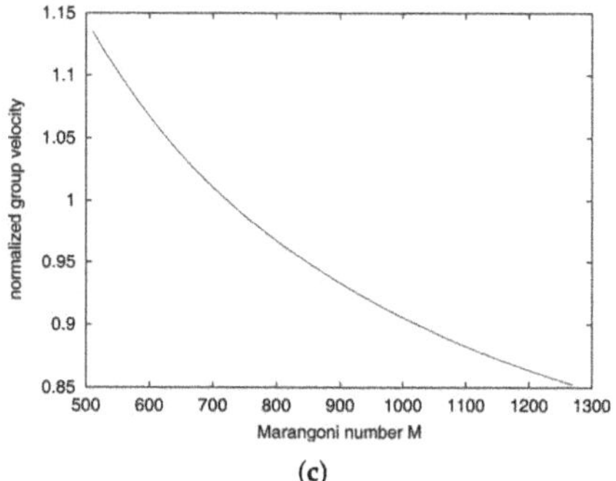

(c)

Figure 4. (**a**) wavenumber k_{max}; (**b**) growth rate $Re\{\omega_{max}\}$; (**c**) group velocity $Im\{\omega\}/k$ reduced by long wave phase velocity (26) for which $\Delta = 0$ and $Re\{\omega_{max}\} > Re\{\omega\}$ for all k at the parametric values $\{K = 0.0001, Pr = 1.001, G = 1\}$ and $M \in \{500, 1250\}$. These curves identify the most dangerous mode (wavelength with most rapid growth or slowest decay). Clearly, there is neutral stability at $M \sim 1100$. The jaggedness of (**a**) reflects the granularity of the discretization in k-M-continuation.

4.2. Critical Parameters via the Neutral Stability Method

Identifying the critical Marangoni number and associated wavenumber and frequency by continuation in k and M for fixed $\{K, Pr, G\}$ is computationally expensive if parameter space is to be mapped. Once a single critical parameter set is known, continuation along the neutral stability surface should be possible. Ref. [14] recognized that the Marangoni number only arises linearly in Equation (20), so that it is possible to solve the tangential stress boundary condition for M and eliminate all quantities α, thus finding the neutral stability curve by the condition that at arbitrary k and with $Re\{\omega\} = 0$, $Im\{M\} = 0$ is imposed by adjusting $s = Im\{\omega\}$. Takashima [15] also used this technique. Neither study, however, identified the long wave stationary branches put forth here. The critical parameters $\{M_{crit}, k_{crit}, s_{crit}\}$ are then found by minimizing M over k. In this paper, the relation $\Delta = 0$ was solved for M, and neutral stability was found by using Newton's method to adjust s to achieve $Im\{M\} = 0$. Use of numerical root finding methods for s such that $Im\{M\} = 0$ are faster than those that seek values of s such that $\Delta = 0$.

An algorithm for parameter space continuation in $\{K, Pr, G\}$ for $\{M_{crit}, k_{crit}, s_{crit}\}$ by traversing the neutral stability surface was developed. The essential steps are to find the M-k and s-k neutral curves by computing a neighborhood in k surrounding an estimated critical point $\{M, k, s\}$ from adjacent values of $\{K, Pr, G\}$. Subsequently, Newton's method is used to find the minimum of the M-k neutral curve. This is illustrated in Figure 5 for the parameters $Pr = 1.001$, $K = 0.0001$ and $G = 1$. The crossover wavenumber $M - k$ neutral stability curve, i.e., that for which $\Re(\omega) = 0$, is plotted, along with a curve showing the functional dependence of the frequency $s = Im(\omega)$ on the wavenumber k. A space of twenty values of each of $\{K, Pr, G\}$ was mapped by this method. Table 1 exhibits a sampling of the database built up from Newton continuation in $\{K, Pr, G\}$. The range of parameters was selected to represent some physically realizable liquids (small capillary number and O(1) Prandtl number) and layer depths (microgravity to terrestrial gravity). From Table 1, it is clear that values of $M_{crit} \sim 10^4$–10^5 are achievable under combinations of small capillary number and thin layers in liquids. Critical wavenumbers k_{crit} are typically of intermediate scale, neither long (<0.1, say) nor short (>10, say), but usually about $k_{crit} \sim 0.25$.

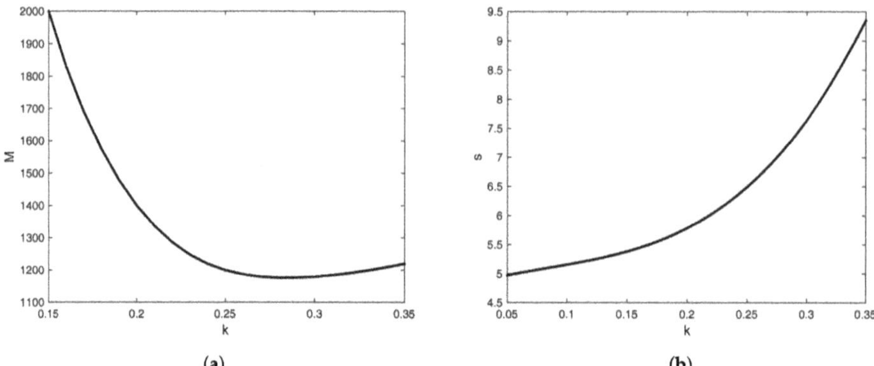

(a) **(b)**

Figure 5. Magnitude of velocity field on $z = 0$ mm plane: (**a**) crossover wavenumber k for the M-k neutral stability curve ($Re\{\omega\} = 0$) for the parameters $\mathrm{Pr} = 1.001, K = 0.0001, G = 1$. M_{crit} is found from the global minimum of the neutral curve and k_{crit} is the associated wavenumber; (**b**) frequency $s = Im\{\omega\}$.

Table 1. Critical values $\{M_{crit}, k_{crit}, s_{crit}\}$ for fixed $\{\mathrm{Pr}, G, K\}$ found by the neutral stability/minimization algorithm and parameter space continuation within the database. Samples were tested against the dispersion relation method and were found to be in agreement and global maximum of growth rate versus k. The table shows only selected values; the database is $\{\mathrm{Pr} \in [1., 10.]\} \times \{K \in [10^{-6}, 10^{-2}]\} \times \{G \in [10^{0}, 10^{5}]\}$. There are twenty values for each parameter, spaced exponentially.

K	Pr	G	k_{crit}	s_{crit}	M_{crit}
1.0×10^{-2}	1.001	1.000	0.2820	6.664	1.032×10^{3}
1.0×10^{-2}	1.001	1.833	0.2821	6.674	1.034×10^{3}
1.0×10^{-2}	1.001	3.360	0.2823	6.692	1.036×10^{3}
1.0×10^{-2}	1.001	6.160	0.2826	6.724	1.041×10^{3}
1.0×10^{-2}	1.001	1.129×10^{1}	0.2833	6.783	1.049×10^{3}
1.0×10^{-2}	1.001	2.069×10^{1}	0.2847	6.892	1.064×10^{3}
1.0×10^{-2}	1.001	3.793×10^{1}	0.2877	7.095	1.087×10^{3}
1.0×10^{-2}	1.001	6.952×10^{1}	0.2949	7.479	1.120×10^{3}
1.0×10^{-2}	1.001	1.274×10^{2}	0.3140	8.264	1.152×10^{3}
1.0×10^{-2}	1.001	2.336×10^{2}	0.3736	10.16	1.130×10^{3}
1.0×10^{-2}	1.001	4.281×10^{2}	0.5052	14.17	1.008×10^{3}
1.0×10^{-2}	1.001	7.848×10^{2}	0.5703	18.22	1.042×10^{3}
1.0×10^{-2}	1.001	1.438×10^{3}	0.5501	21.89	1.334×10^{3}
1.0×10^{-2}	1.001	2.637×10^{3}	0.4826	25.08	1.979×10^{3}
1.0×10^{-2}	1.001	4.833×10^{3}	0.3936	27.34	3.224×10^{3}
1.0×10^{-2}	1.001	8.859×10^{3}	0.3048	28.57	5.548×10^{3}
1.0×10^{-2}	1.001	1.624×10^{4}	0.2296	29.13	9.835×10^{3}
1.0×10^{-2}	1.001	2.976×10^{4}	0.1710	29.38	1.771×10^{4}
1.0×10^{-2}	10.	1.0×10^{4}	0.2105	210.3	1.693×10^{5}
8.859×10^{-4}	10.	1.0×10^{4}	0.2102	210.0	1.694×10^{5}
7.848×10^{-5}	10.	1.0×10^{4}	0.2069	207.4	1.702×10^{5}
1.0×10^{-6}	10.	1.0	0.1563	89.24	6.994×10^{4}
1.0×10^{-6}	10.	1.0×10^{4}	0.1403	156.2	2.096×10^{5}

Figure 4 is suggestive that the group velocity s/k of the critical mode might actually be well predicted by the long wave formula (20) for the phase velocity. The scatter plot in Figure 6 tests this hypothesis by crudely applying it to the entire mapped set of 8000 critical points. The critical group velocity, $c_0 = \sqrt{(G+M)Pr}$, is plotted against the longwave phase velocity, s_{crit}/k_{crit}, for each of the 8000 critical parameter sets considered. The plot suggests that c_0 is the correct order of magnitude for the group velocity of the critical mode regardless of the parameter set in the range mapped. Furthermore, it is observed that the magnitude of the group velocity is constrained to be within $\frac{1}{2}$ to $\frac{2}{3}$ of the longwave phase velocity.

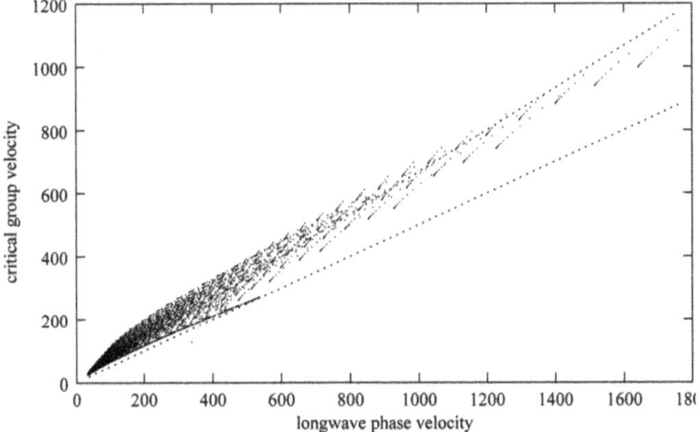

Figure 6. Scatter plot of $c_0 = \sqrt{(G+M)\,Pr}$ versus s_{crit}/k_{crit} for the 8000 critical parameter sets mapped. This suggests that c_0 is the correct order of magnitude and a fair approximation for the group velocity of the critical mode regardless of the parameter set in the range mapped. Dotted lines have slope $\frac{2}{3}$ and $\frac{1}{2}$.

5. Conclusions

An analytical study based on linear theory has been presented that considers the development of instabilities that arise when a thin layer of a stably stratified fluid is heated from above for the Marangoni–Bénard problem. Particular focus was directed towards the value of the critical Marangoni number and to the influence of the lower boundary condition.

The stress-free model modifies the predicted critical Marangoni number to depend on the Galileo number (gravity effects), but reduces to $M_{crit} = 12$ under microgravity. It was not possible to find the cut-off wavenumber analytically as it requires solution of the dispersion relation to $O(k^2)$ in small wavenumber k, but numerical solutions of the secular equation indicates, for all parameters tested, that $k_{cutoff} < \sim 0.05$. This quantitative verification of the assumptions of KdV–KSV theory [8] under the stress-free boundary conditions leads one to conclude that any deficiencies of the theory must lie in the lower boundary condition. The analytic study of the long wave solutions to the secular equation for the no-slip model led to the surprising discovery that there are no infinitely long propagating modes. The three modes identified as $k \to 0$ are a neutral pure temperature mode, a damped viscous (pure velocity) mode, and a damped thermal (pure temperature) mode. The temperature and viscous modes both decay with increasing k, but the thermal mode increases in growth rate with small k. The prediction of an intersection point is computed by Newton continuation of the nonlinear solution of the secular equation. At the intersection of the two modes, the crossover point, there is co-existence of two propagating modes with the two stationary long modes. At higher wavenumbers, only the propagating modes exist. These modes are mapped by parameter space continuation in wavenumber, and a parametric study by continuation in the physical parameter space is made. Further developments in this research area could be facilitated by full numerical simulations.

Acknowledgments: This work was supported by funding from the Engineering and Physical Sciences Research Council (EPSRC) grant EP/I019790/1 on Microbubble cloud generation from fluidic oscillation: underpinning fluid dynamics. Funding was provided from the University of Sheffield for covering the costs to publish in open access.

Author Contributions: William B. Zimmerman and Julia M. Rees contributed equally to the theory and computation.

Conflicts of Interest: The authors declare no conflict of interest. The funding sponsors had no role in the design of the study; in the collection, analyses, or interpretation of data; in the writing of the manuscript, and in the decision to publish the results.

Abbreviations

The following abbreviations are used in this manuscript:

KdV–KSV equation	Korteweg–de Vries–Kuramoto–Sivashinsky–Velarde equation
LST	Linear stability theory

References

1. Velarde, M.G.; Normand, C. Convection. *Sci. Am.* **1980**, *243*, 92–108.
2. Sternling, C.V.; Scriven, L.E. Interfacial turbulence: Hydrodynamic instability and the Marangoni effect. *Am. Inst. Chem. Eng. J.* **1959**, *5*, 514–523.
3. Smith, K.A. On convective instability induced by surface-tension gradients. *J. Fluid Mech.* **1966**, *24*, 401–414.
4. Velarde, M.G.; Nepomnyashchy, A.A.; Hennenberg, M. Onset of oscillatory interfacial instability and wave motions in Bénard layers. *Adv. Appl. Mech.* **2001**, *37*, 167–238.
5. MacInnes, J.M.; Pitt, M.J.; Priestman, G.H.; Allen, R.W.K. Analysis of two-phase contacting in a rotating spiral channel. *Chem. Eng. Sci.* **2012**, *69*, 304–315.
6. Linde, H.; Schwarz, E.; Groger. K. Zum aufreten des oszillatorishen regimes der Marangoni-instabilitat beim Stoffubergang. *Chem. Eng. Sci.* **1967**, *22*, 823–836.
7. Linde, H.; Chu, X.; Velarde, M.G. Oblique and head-on collisions of solitary waves in Marangoni–Bénard convection. *Phys. Fluids* **1993**, *5*, 1068–1070.
8. Nepomnyashchy, A.A.; Velarde, M.G. A three-dimensional description of solitary waves and their interaction in Marangoni-Benard layers. *Phys. Fluids* **1994**, *6*, 187–197.
9. Zimmerman, W.B.; Rees, J.M.; Hewakandamby, B.N. Numerical analysis of solutocapillary Marangoni-induced interfacial waves. *Adv. Colloid Interface. Sci.* **2007**, *134-135*, 346–359.
10. Rees, J.M.; Zimmerman, W.B. An intermediate wavelength, weakly nonlinear theory for the evolution of capillary gravity waves. *Wave Motion* **2011**, *48*, 707–716.
11. Davis, S.H.; Homsy, G.M. Energy stability theory for free-surface problems: Buoyancy-thermocapillary layers. *J. Fluid Mech.* **1980**, *98*, 527–553.
12. Lighthill, J. *Waves in Fluids*; Cambridge University Press: Cambridge, UK, 1978.
13. Joseph, D.D.; Saut, J.C. Short-wave instabilities and ill-posed initial-value problems. *Theor. Comput. Fluid Dyn.* **1990**, *1*, 191–227.
14. Reichenbach, J.; Linde, H. Linear perturbation analysis of surface-tension-driven convection at a plane interface (Marangoni instability). *J. Colloid Interface Sci.* **1981**, *84*, 433–443.
15. Takashima, M. Surface tension driven instability in a horizontal liquid layer with a deformable free surface. II. Overstability. *J. Phys. Soc. Jpn.* **1981**, *50*, 2751–2756.

© 2017 by the authors. Licensee MDPI, Basel, Switzerland. This article is an open access article distributed under the terms and conditions of the Creative Commons Attribution (CC BY) license (http://creativecommons.org/licenses/by/4.0/).

Article

Heat Transfer Investigation of the Unsteady Thin Film Flow of Williamson Fluid Past an Inclined and Oscillating Moving Plate

Taza Gul [1,*], Abdul Samad Khan [2], Saeed Islam [1], Aisha M. Alqahtani [3], Ilyas Khan [4], Ali Saleh Alshomrani [5], Abdullah K. Alzahrani [5] and Muradullah [6]

1 Department of Mathematics, Abdul Wali Khan University, Mardan 32300, KP, Pakistan; saeedislam@awkum.edu.pk
2 Department of Mathematics, Bacha Khan University, Charsadda 244420, KP, Pakistan; abdulsamadkhan17@yahoo.com
3 Department of Mathematics, Princess Nourah bint Abdulrahman University, Riyadh 11564, Saudi Arabia; alqahtani@pnu.edu.sa
4 Basic Engineering Department, College of Engineering, Majmaah University, Majmaah 11952, Saudi Arabia; ilyaskhanqau@yahoo.com
5 Department of Mathematics, King Abdul Aziz University, Jeddah 21577, Saudi Arabia; aszalshomrani@kau.edu.sa (A.S.A.); akalzahrani@kau.edu.sa (A.K.A.)
6 Department of Mathematics, Islamia College, Peshawar 25000, Pakistan; muradullah90@yahoo.com
* Correspondence: tazagulsafil@yahoo.com; Tel.: +92-331-917-1160

Academic Editors: Artur J. Jaworski and Yulong Ding
Received: 30 January 2017; Accepted: 4 April 2017; Published: 7 April 2017

Abstract: This investigation aims at analyzing the thin film flow passed over an inclined moving plate. The differential type non-Newtonian fluid of Williamson has been used as a base fluid in its unsteady state. The physical configuration of the oscillatory flow pattern has been demonstrated and especial attention has been paid to the oscillatory phenomena. The shear stresses have been combined with the energy equation. The uniform magnetic field has been applied perpendicularly to the flow field. The principal equations for fluid motion and temperature profiles have been modeled and simplified in the form of non-linear partial differential equations. The non-linear differential equations have been solved with the help of a powerful analytical technique known as Optimal Homotopy Asymptotic Method (OHAM). This method contains unknown convergence controlling parameters $C_1, C_2, C_3, \ldots$ which results in more efficient and fast convergence as compared to other analytical techniques. The OHAM results have been verified by using a second method known as Adomian Decomposition Method (ADM). The closed agreement of these two methods and the fast convergence of OHAM has been shown graphically and numerically. The comparison of the present work and published work has also been equated graphically and tabulated with absolute error. Moreover, the effect of important physical parameters like magnetic parameter M, gravitational parameter m, Oscillating parameter ω, Eckert number Ec and Williamson number We have also been derived and discussed in this article.

Keywords: time dependent thin film fluid flow; Williamson fluid; oscillating inclined plane; heat transfer and magnetic field; Optimal Homotopy Asymptotic Method (OHAM)

1. Introduction

Thin-film flows have various applications in the fields of engineering and industry. In chemical engineering, thin film layers have an effective role in the developments of thin-film reactors, distillation columns, condensers and evaporators. The huge advantages of a thin layer depends on its minute thickness, which provides its flow through micro channels. Thin film layers have vital roles in physical engineering and have broad applications in providing cooling methods for nanotechnology through heat sinks. In geophysical engineering, a lot of problems emerge related to thin film flows such as mudslides, debris flows and lava [1,2]. In the available literature, studies of thin film flows are mostly related to Newtonian fluids and very little work has been done related to non-Newtonian fluids. These too are of great importance, as shear-thickening and pseudoplastic or shear-thinning fluids are considered to be in the class of non-Newtonian fluids. The studies of thin film flow of shear-thinning fluids have vast importance in the industries involving photographic films and the extrusion of polymer sheets, etc. The behavior of pseudoplastic fluids has been explained comprehensively in the Ostwald de Waele power law, Cross, Carreaus and Ellis models, but little attention has been given to the Williamson fluid model. The flow of pseudoplastic fluids has been experimentally described by Williamson [3], with verified results. Dapra and Scarpi [4] have discussed the analytical solution of Williamson fluid by using the perturbation technique. Most of the problems related to Williamson fluid have been modeled in steady state. Very little work has been done on unsteady non-Newtonian fluids. Gamal and Rahman [5] have discussed the unsteady magnetohydrodynamic flow of non-Newtonian fluids obeying the power law model. Khan [6] has discussed Williamson fluid flow using scaling transformation. Hayat et al. [7] have studied the Williamson fluid flow on a stretching cylinder with thermal radiations and Ohmic dissipation. Nadeem et al. [8] have discussed the Williamson fluid flow on a stretching surface. They studied the effect of physical parameters involved in the problem. Waris et al. [9] have investigated the nanofluid study of the thin film Williamson fluid flow on a stretching surface. Abdollahzadeh Jamalabadi et al. [10] have examined the numerical simulation of Williamson fluid flow amongst binary parallel vertical Walls with slip effects. Thin film nanofluid spray on the stretching cylinder surface has been examined by Noor et al. [11]. The spray rate under the applied pressure force has been scrutinized in their study. Gul et al. [12] have discussed the variable properties of thin layer third order differential type fluid. Miladinova et al. [13] and Siddiqui et al. [14] have studied the thin film flow of non-Newtonian fluids using different geometries. The unsteady fluid problems have been evaluated in the study of Fetecau [15]. The similar study of fluid flows in its unsteady state can also be seen in [16–19].

The fluid flow with oscillatory physical conditions has important applications, such as those that Fatimah and Jaworski [20] have examined in their experimental analysis of the friction factor correlation for a regenerator using oscillatory phenomena in a thermo acoustic system. Huang et al. [21] have experimentally studied the empirical correlation for finned heat exchangers using oscillating analysis. The thin film flow on an oscillatory belt in the field of mechanical engineering has been studied by Gul et al. [22]. Heat transmission in fluid flows has much importance in basic engineering. Still, a lot of work remains to be done regarding oscillatory flow problems. Various applications related to Stirling machines, heat exchangers, pulsed-tube coolers in cryogenics and wire coating have been explored by Shah et al. [23]. Gravity-driven flows may occur in various natural and industrial events, such as the lifting and drainage of fluid flows on a vertical and inclined oscillating plane, as studied by Yongqi et al. [24] and Gul et al. [25–27]. Ellahi et al. [28] have studied the oscillating effect of nano-Ferroliquid on a stretchable rotating disk.

For the solution of real world problems, researchers and engineers use numerical [29,30] and analytical [31–36] techniques to find the approximate solution of modelled problems. This recent work has been handled analytically using the technique known as Optimal Homotpy Asymptotic Method (OHAM). OHAM is one of the powerful analytical techniques to handle non-linear problems. Marinca and Herisanu [37–39] are the founders of OHAM and they applied this method to the problems of thin film and steady flows. Nowadays, most researchers use OHAM for the solution of nonlinear Ordinary Differential Equations (ODEs)and Partial Differential Equations (PDEs). The uses of this method for high non-linear problems can also be seen in the work of Mabood et al. [40].

Considering the above expression, the aim of the current work is to study the thin film flow of Williamson fluid on an oscillating and translated inclined plane in the presence of magnetic and temperature fields using OHAM. The physical study of the magnetic field, gravity term, oscillation parameter and Eckert number is investigated and discussed in this work.

2. Materials and Methods

Consider a flow of thin film Williamson fluid on an inclined plate. The plate is oscillating and moving with a uniform velocity U at time $t = 0^+$. The thickness of the liquid film is considered uniform and equal to δ. The gravity force opposes the fluid motion and tries to make the fluid film drain down the plate. A uniform magnetic field is applied transversely to the translating inclined plate. Furthermore, during fluid motion, constant temperature filed is also considered. The flow is assumed to be unsteady and laminar after a small distance above the film surface.

The physical conditions of the unsteady problem are used as:

$$u(0,t) = U(1+\cos\omega t), \frac{\partial u(\delta,t)}{\partial y} = 0, \tag{1}$$

$$\Theta(0,t) = \Theta_0, \Theta(\delta,t) = \Theta_1, \tag{2}$$

where ω is the frequency of the oscillating plate and δ is the thickness of the thin film.

The Lorentz force $\mathbf{J} \times \mathbf{B}$ for the uniform Magnetic field is defined as:

$$\mathbf{J} \times \mathbf{B} = \left[0, -\sigma B_0^2 u(y,t), 0\right], \tag{3}$$

where σ is the electrical conductivity of the fluid.

The governing equations take the form as:

$$\nabla.\mathbf{u} = 0, \tag{4}$$

$$\rho\frac{d\mathbf{u}}{dt} = -\nabla p + \rho\mathbf{g}\sin(\Theta) + \nabla.\mathbf{T} + \mathbf{J} \times \mathbf{B}, \tag{5}$$

$$\rho C_p\frac{D\Theta}{Dt} = k\nabla^2\Theta + tr(\mathbf{T}.\mathbf{L}), \tag{6}$$

where $\mathbf{u}$ is the velocity vector, ρ is the fluid density, $\mathbf{g}$ is the gravity force, p is the pressure term, Θ is the temperature field, k defines thermal conductivity, C_p is the specific heat and $\mathbf{L} = \nabla\mathbf{u}$.

The extra stress tensor $\mathbf{T}$ for the Williamson fluid is defined as in [4,8]:

$$\mathbf{T} = \left[\mu_\infty + (\mu_0 + \mu_\infty)\left(1 - \Gamma\dot{\gamma}\right)^{-1}\right]\dot{\gamma}, \tag{7}$$

where Γ is the time constant, μ_0 is the zero viscosity, μ_∞ is the infinite viscosity, and $\dot{\gamma}$ is defined as:

$$\dot{\gamma} = \sqrt{\frac{1}{2}\sum_i\sum_j \gamma_{ij}\gamma_{ji}} = \sqrt{\frac{1}{2}\mathbf{\Pi}}, \tag{8}$$

where $\mathbf{\Pi} = trace(A_1)^2 = 2\left(\frac{\partial u}{\partial y}\right)^2$ is the second invariant strain tensor derived as in [4,8]. We consider the constitutive Equation (7) in which $\mu_\infty = 0$ and $\Gamma\dot{\gamma} < 1$. The component of the extra stress tensor, therefore, can be written as:

$$\mathbf{T} = \mu_0\left[\left(1 - \Gamma\dot{\gamma}\right)^{-1}\right]\dot{\gamma} = \mu_0\left[\left(1 + \Gamma\dot{\gamma}\right)\right]\dot{\gamma}, \tag{9}$$

where $\dot{\gamma} = \frac{\partial u}{\partial y}$ is the shear rate and Γ is the time constant. The unsteady unidirectional velocity field and temperature profile is defined as:

$$\mathbf{u} = [u(y,t), 0, 0], \ \Theta = \Theta(y,t), \tag{10}$$

When (9) and (10) are used in (4)–(6), the continuity equation is identically satisfied and the momentum and energy equations without dynamic pressure are reduced to:

$$\rho\frac{\partial u}{\partial t}=\mu_0\frac{\partial^2 u}{\partial y^2}+\mu_0\Gamma\frac{\partial}{\partial y}\left(\frac{\partial u}{\partial y}\right)^2-\rho g\sin(\Theta)-\sigma B_0^2 u, \tag{11}$$

$$\rho c_p\left(\frac{\partial\Theta}{\partial t}\right)=k\frac{\partial^2\Theta}{\partial y^2}+\mu_0\left(\frac{\partial u}{\partial y}\right)^2+\mu_0\Gamma\left(\frac{\partial u}{\partial y}\right)^3, \tag{12}$$

Introduce the following dimensionless parameters:

$$\begin{gathered}\overline{u}=\tfrac{u}{U},\ \overline{y}=\tfrac{x}{\delta},\ \overline{t}=\tfrac{\mu t}{\rho\delta^2},\ \overline{\Theta}=\tfrac{\Theta-\Theta_0}{\Theta_1-\Theta_0},\ \overline{\omega}=\tfrac{\omega\delta^2\rho}{\mu},\ We=\tfrac{\Gamma U}{\delta},\\ P_r=\tfrac{\mu C_p}{k},\ E_c=\tfrac{\mu U^2}{k(\Theta_1-\Theta_0)},\ M=\tfrac{\sigma B_0^2\delta^2}{\mu_0},\ m=\tfrac{\delta^2\rho g\sin\Theta}{\mu U}.\end{gathered} \tag{13}$$

Using the above dimensionless parameters from Equation (13) in the governing partial differential Equations (11) and (12) and in the boundary conditions (1,2), we get:

$$\frac{\partial u}{\partial t}=\frac{\partial^2 u}{\partial y^2}+W_e\frac{\partial}{\partial y}\left(\frac{\partial u}{\partial y}\right)^2-m-Mu, \tag{14}$$

$$Pr\left(\frac{\partial\Theta}{\partial t}\right)=\frac{\partial^2\Theta}{\partial y^2}+E_c\left(\frac{\partial u}{\partial y}\right)^2+W_eE_c\left(\frac{\partial u}{\partial y}\right)^3, \tag{15}$$

$$u(0,t)=(1+\cos\omega t),\ \frac{\partial u(1,t)}{\partial t}=0, \tag{16}$$

$$\Theta(0,t)=0,\ \Theta(1,t)=1. \tag{17}$$

where m, is the gravitational parameter, W_e is the Williamson parameter, M is the magnetic parameter, Pr is the Prandtl number, E_c is the Eckert number, ω is the oscillating parameter and t is the time parameter.

Solution by OHAM

The boundary value problem is considered to analyze the OHAM method as:

$$L(u(y,t))+N(u(y,t))+g(u(y,t))=0, B(u)=0. \tag{18}$$

Here L denotes the linear operator of the differential equation, N is used as a non-linear operator, the independent variable is denoted by y, g is a source term and B is a boundary operator. According to OHAM, the homotopy is constructed as:

$$[1-q][L\psi(y,t,q)+g(y,t)]-H(q)[L\psi(y,t,q)+g\psi(y,t,q)+N\psi(y,t,q)]=0, \tag{19}$$

where $q\in[0,1]$ is an embedding parameter, $H(q)=qc_1+q^2c_2+...m$, is an auxiliary function, c_1,c_2 are the convergence controlling parameters and $\psi(x,t,q)$ is an unknown function. Obviously, when $q=0$ and $q=1$, it holds that:

$$\psi(y,t,q)=u_0(y,t),\ \psi(y,t,1)=u(y,t), \tag{20}$$

$$\psi(y,t,q,c_i)=u_0(y,t)+\sum_{k\geq 1}u_k(y,t,c_i)q^k, i=1,2,3...,m, \tag{21}$$

Implanting Equation (20) in Equation (21), accumulating the similar powers of q and comparing each coefficient of q to zero, the non-linear PDE are solved with the given boundary conditions to get $u_0(y,t)$, $u_1(y,t)$, $u_2(y,t)$.

The general solution of Equation (21) can be written as:

$$u^m = u_0(y,t) + \sum_{k=1} u_k(y,t,c_i), \tag{22}$$

The coefficients $c_1, c_2, c_3, ..., c_m$ are the functions of y.

Inserting Equation (22) in Equation (18), the residual is obtained as:

$$R(y,t,c_i) = L(u^m(y,t,c_i)) + g(y,t) + N(u^m(y,t,c_i)), \tag{23}$$

Frequently methods like Ritz Method, Galerkin's Method, Collocation Method and Method of Least Squares are used to find the optimal values of c_i, $i = 1,2,3,4....$ The current problem is solved through the Method of Least Squares as given below:

$$J(c_1,c_2,c_3,...,c_m) = \int_a^b R^2(y,t,c_1,c_2,c_3,...,c_m)dy, \tag{24}$$

where a and b are constants and selected from the domain of the problem.

The controlling convergence parameters $(c_1, c_2, c_3, ..., c_m)$ can be obtained from:

$$\frac{\partial J(c_1,c_2,...,c_m)}{\partial c_1} = \frac{\partial J(c_1,c_2,...,c_m)}{\partial c_2} = ... = \frac{\partial J(c_1,c_2,...,c_m)}{\partial c_m} = 0 \tag{25}$$

Finally, from these controlling convergence parameters, the approximate solution is determined.

The homotopy equation constructed in Equation (18) is applied to Equations (14) and (15) and the like powers of q are equated as:

$$q^o : \frac{\partial^2 u_0}{\partial y^2} = m, \tag{26}$$

$$\frac{\partial^2 \Theta_0}{\partial y^2} = 0, \tag{27}$$

$$q^1 : \frac{\partial^2 u_1}{\partial y^2} = -m + mc_1 - Mc_1v_0 + c_1\frac{\partial v_0}{\partial t} + \frac{\partial^2 v_0}{\partial y^2} - c_1\frac{\partial^2 v_0}{\partial y^2} + 2c_1W_e\frac{\partial v_0}{\partial y}\left(\frac{\partial^2 v_0}{\partial y^2}\right), \tag{28}$$

$$\frac{\partial^2 \Theta_1}{\partial y^2} = \Pr c_3\frac{\partial \Theta_0}{\partial t} + E_c c_3\left(\frac{\partial v_0}{\partial y}\right)^2 - E_c W_e c_3\left(\frac{\partial v_0}{\partial y}\right)^3 + \frac{\partial^2 \Theta_0}{\partial y^2} - c_3\frac{\partial^2 \Theta_0}{\partial y^2}, \tag{29}$$

$$\begin{aligned} q^2 : \frac{\partial^2 u_2}{\partial y^2} = & -mc_2 - Mc_2u_0 - c_2\frac{\partial u_0}{\partial t} - Mc_1u_1 - c_1\frac{\partial u_1}{\partial t} + c_2\frac{\partial^2 u_0}{\partial y^2} - 2W_ec_2\left(\frac{\partial u_0}{\partial y}\right)\left(\frac{\partial^2 u_0}{\partial y^2}\right) - \\ & 2W_ec_1\left(\frac{\partial u_1}{\partial y}\right)\left(\frac{\partial^2 u_0}{\partial y^2}\right) - (1-c_1)\frac{\partial^2 u_1}{\partial y^2} - 2W_ec_1\left(\frac{\partial u_0}{\partial y}\right)\left(\frac{\partial^2 u_1}{\partial y^2}\right), \end{aligned} \tag{30}$$

$$\begin{aligned} \frac{\partial^2 \Theta_2}{\partial y^2} = & -\Pr c_4\frac{\partial \Theta_0}{\partial t} - \Pr c_3\frac{\partial \Theta_1}{\partial t} + E_c c_4\left(\frac{\partial u_0}{\partial y}\right)^2 + W_eE_c(c_4+c_3)\frac{\partial u_0}{\partial y}\left(\frac{\partial u_0}{\partial y}\right)^2 + \\ & 2E_c c_3\frac{\partial u_0}{\partial y}\frac{\partial u_1}{\partial y} + c_4\frac{\partial^2 \Theta_0}{\partial y^2} + (1+c_3)\frac{\partial^2 \Theta_1}{\partial y^2}, \end{aligned} \tag{31}$$

The zeroth, first and second components solution for the velocity and temperature fields are obtained from Equations (26)–(31) using the boundary conditions from Equations (16) and (17), respectively.

$$u_0(y,t) = 1 + \mathrm{Cos}(t\omega) - my + \frac{m}{2}y^2, \tag{32}$$

$$\Theta_0(y,t) = y, \tag{33}$$

$$u_1(y,t,c_1) = c_1\left[(a_0 + a_1\mathrm{Cos}(\omega t))y - (a_2 + a_3\mathrm{Cos}(\omega t))y^2 + \left(a_4 + \tfrac{M}{6}\mathrm{Cos}(\omega t)\right)y^3 - \left(c_1\tfrac{mM}{24}\right)y^4\right], \tag{34}$$

$$u_2(y,t,c_1) = \left[\begin{array}{l} (c_1 b_0 + c_1^2 b_1 + c_1 b_2 \mathrm{Cos}(\omega t) + c_1^2 b_3 \mathrm{Cos}(\omega t) + c_1^2 b_4 \mathrm{Cos}(2\omega t) + c_2 b_5 + c_2 b_6 \mathrm{Cos}(\omega t))y + \\ (c_1 b_7 + c_2 b_8 \mathrm{Cos}(\omega t) + c_1^2 b_9 + c_1^2 b_{10} \mathrm{Cos}(\omega t) + c_1^2 b_{11} \mathrm{Cos}(2\omega t) + c_2 b_{12} - c_2 m w_e \mathrm{Cos}(\omega t))y^2 + \\ \left(c_1 b_{13} + c_1 \frac{M}{6} \mathrm{Cos}(\omega t) + c_1^2 b_{14} + c_1^2 b_{15} \mathrm{Cos}(\omega t) - c_1^2 \frac{M w_e}{6} \mathrm{Cos}(2\omega t) + c_2 b_{16} + c_2 \frac{M}{6} \mathrm{Cos}(\omega t)\right) y^3 \\ \left(-c_1 \frac{mM}{24} + c_1^2 b_{17} + c_1^2 b_{18} \mathrm{Cos}(\omega t)\right) y^4 + \left(\left(\frac{c_1^2 M}{12}\right) b_{19}\right) y^5 + \left(c_1^2 \frac{mM^2}{720}\right) y^6 \end{array} \right], \tag{35}$$

$$u(y,t,c_1) = 1 + \mathrm{Cos}(\omega t) + \left[\begin{array}{l} \left(\begin{array}{l} -1 - \frac{m}{2} - \mathrm{Cos}(\omega t) + c_1 d_0 + c_1 d_1 \mathrm{Cos}(\omega t) + c_1^2 d_2 + c_1^2 d_3 \mathrm{Cos}(\omega t) + c_1^2 d_4 \mathrm{Cos}(2\omega t) + c_2 d_5 + \\ c_2 d_6 \mathrm{Cos}(\omega t) \end{array} \right) y + \\ \left(\frac{m}{2} + c_1 d_7 + c_1 d_8 \mathrm{Cos}(\omega t) + c_1^2 d_9 + c_1^2 d_{10} \mathrm{Cos}(\omega t) + c_1^2 d_{11} \partial \mathrm{Cos}(2\omega t) + c_2 d_{12} + c_2 d_{13} \mathrm{Cos}(\omega t)\right) y^2 + \\ \left(c_1 d_{14} + c_1 \frac{m}{3} \mathrm{Cos}(\omega t) + c_1^2 d_{15} + c_1^2 d_{16} \mathrm{Cos}(\omega t) - c_1^2 (\frac{M w_e}{6}) \mathrm{Cos}(2\omega t) + c_2 d_{17} + c_2 (\frac{m}{6}) \mathrm{Cos}(\omega t)\right) y^3 + \\ \left((\frac{mM}{12}) c_1 + c_1^2 b_{18} + c_1^2 b_{19} \mathrm{Cos}(\omega t) + (\frac{-mM}{24}) c_2\right) y^4 + \left(c_1^2 d_{20} - (\frac{M^2}{120}) c_1^2 \mathrm{Cos}(\omega t)\right) y^5 + \left((\frac{mM^2}{720}) c_1^2\right) y^6 \end{array} \right], \tag{36}$$

$$\Theta_1(y,t) = \left[\begin{array}{l} (E_c c_3 (e_0 + e_1 \cos(\omega t) + e_2 \cos(2\omega t) - (\frac{w_e}{8}) \cos(3\omega t)))y + \\ (E_c c_3 (e_3 + e_4 \cos(\omega t) + e_5 \cos(2\omega t) + (\frac{w_e}{8}) \cos(3\omega t)))y^2 + \\ (E_c c_3 (e_6 + e_7 \cos(\omega t) - (\frac{m w_e}{4}) \cos(2\omega t)))y^3 + \\ \left(E_c c_3 \left(e_8 + \left(\frac{m^2 w_e}{4}\right) \cos(\omega t)\right)\right) y^4 - \left(E_c c_3 \frac{m^3 w_e}{20}\right) y^5 \end{array} \right], \tag{37}$$

$$\Theta_2(y,t) = \left[\begin{array}{l} \left(E_c (f_0 + f_1 \cos(\omega t)) \cos\left[\frac{\omega t}{2}\right]^4\right) y + \left(E_c (f_2 + f_3 \cos(\omega t)) \cos\left[\frac{\omega t}{2}\right]^4\right) y^2 + \\ \left(E_c (f_4 + 2 M c_1 c_3 w_e \cos(\omega t)) \cos\left[\frac{\omega t}{2}\right]^4\right) y^3 + \left(E_c \left(f_5 - \left(\frac{M c_1 c_3 w_e}{2}\right) \cos(\omega t)\right) \cos\left[\frac{\omega t}{2}\right]^4\right) y^4 \end{array} \right], \tag{38}$$

$$\Theta(y,t) = \left[\begin{array}{l} \left(1 + E_c \left(c_3 g_0 + c_3 g_1 \cos(\omega t) + c_3 g_2 \cos(2\omega t) + g_3 \cos\left[\frac{\omega t}{2}\right]^4 + w_e g_4 \cos\left[\frac{\omega t}{2}\right]^4 \cos(\omega t)\right)\right) y + \\ \left(E_c \left(c_3 g_5 + c_3 g_6 \cos(\omega t) + c_3 g_7 \cos(2\omega t) + g_8 \cos\left[\frac{\omega t}{2}\right]^4 + g_9 \cos\left[\frac{\omega t}{2}\right]^4 \cos(\omega t) + (\frac{c_3 w_e}{8}) \cos(3\omega t)\right)\right) y^2 + \\ \left(E_c c_3 \left(g_{10} + g_{11} \cos(\omega t) + g_{12} \cos\left[\frac{\omega t}{2}\right]^4 - (\frac{m w_e}{4}) \cos(2\omega t) + 2 M c_1 w_e \cos\left[\frac{\omega t}{2}\right]^4 \cos(\omega t)\right)\right) y^3 + \\ \left(E_c c_3 \left(g_{13} + g_{14} \cos\left[\frac{\omega t}{2}\right]^4 + \left(\frac{m^2 w_e}{4}\right) \cos(\omega t) - \left(\frac{M c_1 w_e}{2}\right) \cos\left[\frac{\omega t}{2}\right]^4 \cos(\omega t)\right)\right) y^4 - E_c c_3 \left(\frac{m^3 w_e}{20}\right) y^5 \end{array} \right], \tag{39}$$

3. Basic Idea of Adomian Decomposition Method (ADM)

The Adomian Decomposition Method (ADM) is used to show the unknown function $u(y,t)$ in the form of an infinite series.

$$u(y,t) = \sum_{n=0}^{\infty} u_n(y,t), \tag{40}$$

The components $u_0(y,t), u_1(y,t), u_2(y,t), \ldots..$ of the infinite series can easily be obtained through simple integrals. The summary of the ADM is shown by considering a partial differential equation in an operator form as:

$$L_t u(y,t) + L_y u(y,t) + Ru(y,t) + Nu(y,t) = f(y,t), \tag{41}$$

$$L_y u(y,t) = f(y,t) - L_t u(y,t) - Ru(y,t) - Nu(y,t), \tag{42}$$

where $L_y = \frac{\partial^2}{\partial y^2}$ and $L_t = \frac{\partial}{\partial t}$ are invertible and linear operators, $Ru(y,t)$ is a remaining linear term, $f(y,t)$ is a source term, $Nu(y,t)$ is non-linear part of the partial differential equation and easily expandable in the Adomian polynomials $A_n(y,t)$. After applying the inverse operator L_y^{-1} to both sides of Equation (49), we get:

$$L_y^{-1} L_y u(y,t) = L_y^{-1} f(y,t) - L_y^{-1} L_t u(y,t) - L_y^{-1} Ru(y,t) - L_y^{-1} Nu(y,t), \tag{43}$$

$$u(y,t) = g(y,t) - L_y^{-1} L_t u(y,t) - L_y^{-1} Ru(y,t) - L_y^{-1} Nu(y,t), \tag{44}$$

here, the function $g(y,t)$ represents the terms arising from $L_y^{-1} f(y,t)$ after using the given conditions. $L_y^{-1} = \iint (.) dy dy$ is used as the inverse operator for the second order partial differential equation. Similarly, it is used for the higher order partial differential equation L_y^{-1} and L_y depends on the order of the partial differential equation.

Adomian Decomposition Method defines the series solution $u(y,t)$ as:

$$u(y,t) = \sum_{n=0}^{\infty} u_n(y,t), \tag{45}$$

$$\sum_{n=0}^{\infty} u_n(y,t) = g(y,t) - L_y^{-1} R \sum_{n=0}^{\infty} u_n(y,t) - L_y^{-1} N \sum_{n=0}^{\infty} u_n(y,t), \tag{46}$$

The non-linear term expands in Adomian polynomials as:

$$N \sum_{n=0}^{\infty} u_n(y,t) = \sum_{n=0}^{\infty} A_n, \tag{47}$$

where the components $u_0(y,t), u_1(y,t), u_2(y,t), \ldots..$ are periodically derived as:

$$u_0(y,t) + u_1(y,t) + u_2(y,t) + \ldots.. = g(y,t) - L_y^{-1} R(u_0(y,t) + u_1(y,t) + u_2(y,t) + \ldots) - L_y^{-1}(A_0 + A_1 + \ldots), \tag{48}$$

To determine the series components $u_1(y,t) = -L_y^{-1} R[u_0(y,t)] - L_y^{-1}[A_0]$, it is important to note that ADM suggests that the function $g(y,t)$ actually describes the zeroth component $u_0(y,t)$.

The formal recursive relation is defined as:

$$u_0(y,t) = g(y,t), \tag{49}$$

$$u_1(y,t) = -L_y^{-1} R[u_0(y,t)] - L_y^{-1}[A_0], \tag{50}$$

$$u_2(y,t) = -L_y^{-1} R[u_1(y,t)] - L_y^{-1}[A_1], \tag{51}$$

$$u_3(y,t) = -L_y^{-1} R[u_2(y,t)] - L_y^{-1}[A_2], \tag{52}$$

and so on.

The ADM Solution of the Problem

The inverse operator $L_y^{-1} = \iint (.) dy dy$ is applied on the second order differential Equations (14) and (15) and, using the standard form of ADM, we get:

$$u(y,t) = g(y,t) + M L_y^{-1} u(y,t) + L_y^{-1}\left[\frac{\partial u(y,t)}{\partial t}\right] - W_e L_y^{-1}\left[\frac{\partial}{\partial y}\left(\frac{\partial u(y,t)}{\partial y}\right)^2\right], \tag{53}$$

$$\Theta(y,t) = h(y,t) + \Pr L_y^{-1}\left[\frac{\partial \Theta(y,t)}{\partial t}\right] - Ec L_y^{-1}\left[\left(\frac{\partial u(y,t)}{\partial y}\right)^2 + W_e\left(\frac{\partial u(y,t)}{\partial y}\right)^3\right], \tag{54}$$

To obtain the series solutions of Equations (53) and (54), summation is used as:

$$\sum_{n=0}^{\infty} u_n = g(y,t) + M L_y^{-1}\left[\sum_{n=0}^{\infty} u_n(y,t)\right] + L_y^{-1}\left[\frac{\partial}{\partial t}\sum_{n=0}^{\infty} u_n(y,t)\right] - W_e L_y^{-1}\left[\sum_{n=0}^{\infty} A_n(y,t)\right], \tag{55}$$

$$\begin{aligned} \textstyle\sum_{n=0}^{\infty} \Theta_n(y,t) = h(y,t) + \Pr L_y^{-1}\left[\frac{\partial}{\partial t}\sum_{n=0}^{\infty} \Theta_n(y,t)\right] - \\ Ec L_y^{-1}\left[\sum_{n=0}^{\infty} B_n(y,t)\right] - Ec W_e L_y^{-1}\left[\sum_{n=0}^{\infty} C_n(y,t)\right], \end{aligned} \tag{56}$$

For $n \geq 0$ the Adomian polynomials $A_n(y,t), B_n(y,t)$ and $C_n(y,t)$ from Equations (53) and (54) are defined as:

$$\begin{aligned} \textstyle\sum_{n=0}^{\infty} A_n(y,t) = \sum_{n=0}^{\infty} \frac{\partial}{\partial y}\left(\frac{\partial u_n(y,t)}{\partial y^n}\right)^2, \sum_{n=0}^{\infty} B_n(y,t) = \sum_{n=0}^{\infty}\left(\frac{\partial u_n(y,t)}{\partial y^n}\right)^2, \\ \textstyle\sum_{n=0}^{\infty} C_n(y,t) = \sum_{n=0}^{\infty}\left(\frac{\partial u_n(y,t)}{\partial y^n}\right)^3, \end{aligned} \tag{57}$$

In components form Equations (55) and (56) are derived as:

$$u_0(y,t)+u_1(y,t)+u_2(y,t)+\ldots..=g(y,t)+ML_y^{-1}(u_0(y,t)+u_1(y,t)+u_2(y,t)+\ldots)+ L_y^{-1}\left[\frac{\partial}{\partial t}(u_0(y,t)+u_1(y,t)+u_2(y,t)+\ldots)\right]-W_eL_y^{-1}(A_0(y,t)+A_1(y,t)+A_2(y,t)+\ldots), \quad (58)$$

$$\Theta_0(y,t)+\Theta_1(y,t)+\Theta_2(y,t)+\ldots=h(y,t)+\text{Pr}L_y^{-1}\left[\frac{\partial}{\partial t}(\Theta_0(y,t)+\Theta_1(y,t)+\Theta_2(y,t)+\ldots)\right]- EcL_y^{-1}[(B_0(y,t)+B_1(y,t)+B_2(y,t)+\ldots)-EcW_eL_y^{-1}[(C_0(y,t)+C_1(y,t)+C_2(y,t)+\ldots), \quad (59)$$

Expanding Equations (58) and (59) and comparing both sides for the velocity and temperature fields components, we get:

$$u_0(y,t)=g(y,t)=L_y^{-1}\left(\frac{\partial^2u_0}{\partial y^2}-m\right), \quad (60)$$

$$\Theta_0(y,t)=h(y,t)=L_y^{-1}\left(\frac{\partial^2\Theta_0}{\partial y^2}\right), \quad (61)$$

$$u_1(y,t)=ML_y^{-1}[u_0(y,t)]+L_y^{-1}\left(\frac{\partial u_0(y,t)}{\partial t}\right)-W_eL_y^{-1}[A_0(y,t)], \quad (62)$$

$$\Theta_1(y,t)=\text{Pr}L_y^{-1}\left(\frac{\partial\Theta_0(y,t)}{\partial t}\right)-EcL_y^{-1}[B_0(y,t)]-W_eEcL_y^{-1}[C_0(y,t)], \quad (63)$$

$$u_2(y,t)=ML_y^{-1}[u_1(y,t)]+L_y^{-1}\left(\frac{\partial u_1(y,t)}{\partial t}\right)-W_eL_y^{-1}[A_1(y,t)], \quad (64)$$

$$\Theta_2(y,t)=\text{Pr}L_y^{-1}\left(\frac{\partial\Theta_1(y,t)}{\partial t}\right)-EcL_y^{-1}[B_1(y,t)]-W_eEcL_y^{-1}[C_1(y,t)], \quad (65)$$

Using the physical conditions from Equations (16) and (17) in the Equations (60)–(65), the zeroth and first component solution are obtained as:

$$u_0(y,t)=1+\text{Cos}(\omega t)-my+\left(\frac{m}{2}\right)y^2, \quad (66)$$

$$\Theta_0(y,t)=y, \quad (67)$$

$$u_1(y,t)=\left(m^2W_e+\tfrac{mM}{3}+\omega Sin[\omega t]-Cos[\omega t]-1\right)y+ \tfrac{1}{2}(1+Cos[\omega t]-\omega Sin[\omega t]+2m^2W_e)y^2-\tfrac{1}{6}(mM+2m^2W_e)y^3+\left(\tfrac{mM}{24}\right)y^4, \quad (68)$$

$$\Theta_1(y,t)=\tfrac{m^2Ec}{20}\left[(15-12mW_e)y+(30-30mW_e)y^2+(-20+30mW_e)y^3+(5-15mW_e)y^4+3mW_ey^5\right], \quad (69)$$

The second term solution for velocity field and temperature profile is too bulky, therefore, only graphical representations up to the second order are given.

4. Discussion

Due to the large uses of shear thinning fluids in medicines, especially for the preparation of drugs using oscillatory phenomena to mix the fluid particles to maintain uniformity and other engineering applications, the Williamson fluid is selected from the class of pseudoplastic fluids. Therefore, the recent work has been carried out considering the time-dependent thin film fluid flow of the Williamson fluid model. The study has been carried out on an inclined oscillating plate which is moving with constant velocity U and a constant magnetic field is applied to the plate perpendicularly in the presence of heat transmission. The physical configuration of the problem has been shown in Figure 1. The modelled equations of velocity and temperature fields have been solved analytically by using OHAM. The effect of various embedded parameters has been discussed in Figures 2–10. The physical presentation of the oscillation effect for the velocity profile has been shown in Figure 2. The 3D figure demonstrates that initially the fluid oscillates with the plate jointly and, due to the strong cohesive forces of the non-Newtonian Williamson fluid, its amplitude towards the free surface decreases gradually. Figure 3 describes the 3D presentation of the temperature field. Small amplitude has been observed due to the effect of oscillating shear stresses, which has been combined with the energy equation. Fluid distribution during oscillation ωt in 2D has been shown in Figure 4. This shows that the fluid flow oscillates together with the plate and this oscillation decreases towards the free surface. Physically, the complexity of non-Newtonian behavior and the strong internal forces among the fluid molecules produce friction forces and, as a result, the amplitude of the fluid flow falls towards the free surface. The effect of the temperature field during oscillation has been shown in Figure 5. The amplitude of the flow increases gradually when the temperature increases. In fact, the cohesive forces among the fluid molecules, which keep these molecules closely compacted, reduce and, as a result, the amplitude of the fluid flow rises. The magnetic effect during fluid motion has been shown in Figure 6. Increasing the magnetic parameter results in decreasing the fluid velocity. The reason for this is that the Lorentz force acts against the flow and, as a result, resistance force increases which reduces the flow motion. The effect of the oscillating parameter ω has been revealed in Figure 7. The rising values of the oscillating parameter increase the fluid flow and the effect is similar as in [26,27]. The effect of Williamson number has been shown in Figure 8. Fluid motion reduces with the rise in Williamson number. The reason for this is that the rise in relaxation time causes higher resistance in the fluid flow and, as a result, the velocity field reduces. Figure 9 shows the effect of the gravitational parameter m on the velocity field. Increasing the gravitational parameter m decreases the velocity field. The plate moves in an upward direction and carries a fluid layer in its own direction while the gravitational force acts in the opposite (downward) direction. Therefore, the gravitational force opposes the motion of the fluid and tends to reduce the motion of the fluid particles. Increasing the Eckert number results in an increase in fluid motion because the thermal boundary layer thickness increases by the said increase in the Eckert number. As a result, the cohesive forces decrease which in turn increases the velocity of fluid shown, as shown in Figure 10. Increasing the Williamson number during temperature distribution results in the motion of the fluid increasing as well, as shown in Figure 11. It is noted that the increase is small near the plate and rises more rapidly towards the free surface, in agreeance with [26,27]. The comparison of the present work and published work [22] has been shown in Figures 12 and 13, where the common parameters have been counted and dissimilar parameters have been ignored. In the published work the plate moves vertically while in the present work the plate moves at an incline, with the same physical conditions. The gravitational parameter in the published work is known as stock number S_t and in the present work it has been denoted by m. In Figure 12, the larger values of the second grade parameter α and Williamson number W_e show that the results only match at the boundaries. However, in Figure 13, if we reduce the values of these two parameters, the graphical comparison becomes very close which specifies the validation of this work. The comparison of OHAM and ADM methods for the velocity field and temperature distribution have been shown in Figures 14 and 15. It is clear from Figure 14 that the OHAM and ADM agree with each other initially, and the fast convergence of OHAM solution is clear at the boundary. Similarly, in Figure 15, the boundary conditions are more clearly satisfied by OHAM as compared to ADM. Tables 1 and 2 show the numerical comparison of the present work with published work [22]. The absolute error is larger for higher values of the second grade parameter α and Williamson number We, as shown in Table 1, but this error reduces for small values of these parameters. Tables 3 and 4 demonstrate the agreement of OHAM and ADM for velocity and temperature profiles, respectively, and the closed convergence of OHAM has been obtained.

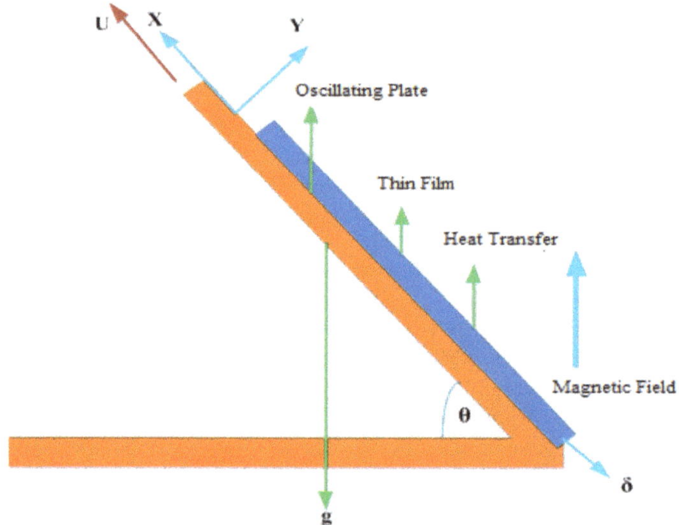

Figure 1. The physical configuration of the problem considering thin film flow of the Williamson fluid passing over an inclined, oscillating, and moving belt.

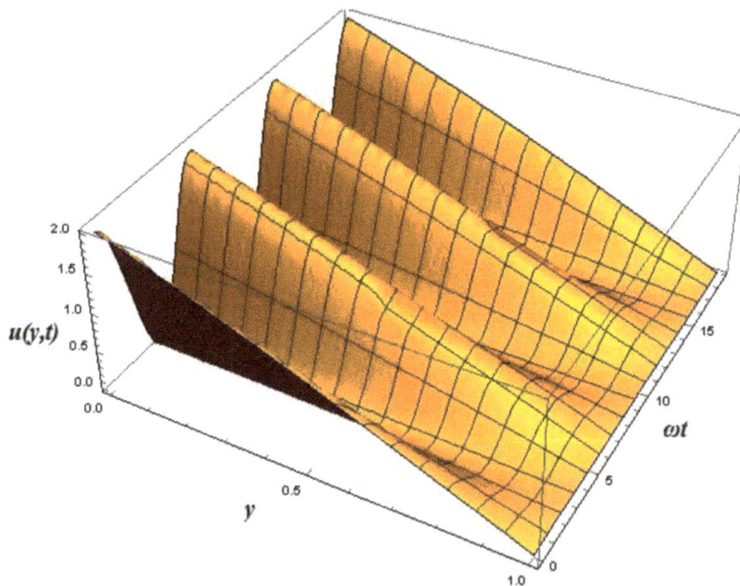

Figure 2. The view of the fluid motion at time level $\omega t \in (0, 6\pi)$ in 3D, when $m = 0.3$, $M = 0.2$, $W_e = 0.1$.

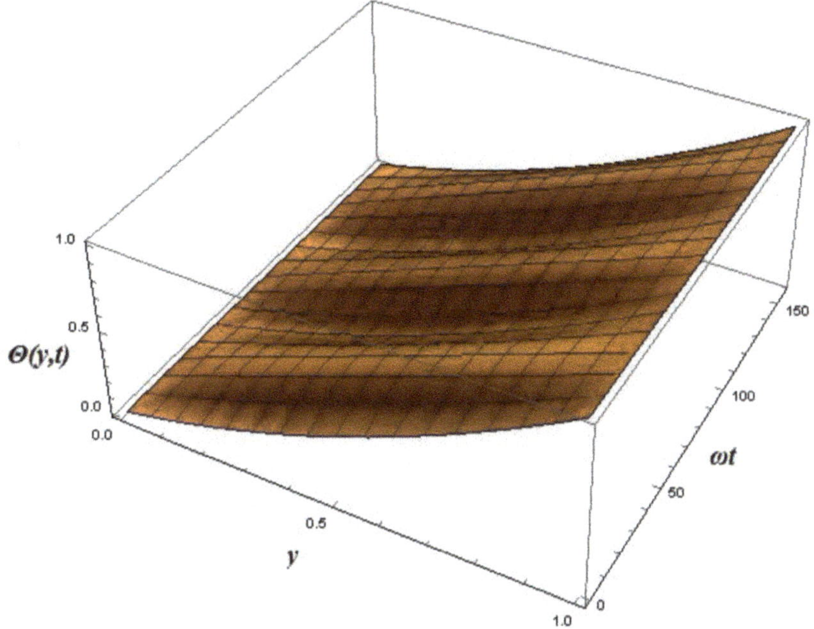

Figure 3. The 3D presentation of temperature distribution at time level $\omega\ t \in (0, 50\pi)$, when $\omega = 0.2$, $m = 0.10$, $M = 0.2$, $W_e = 0.1$, $Ec = 9$.

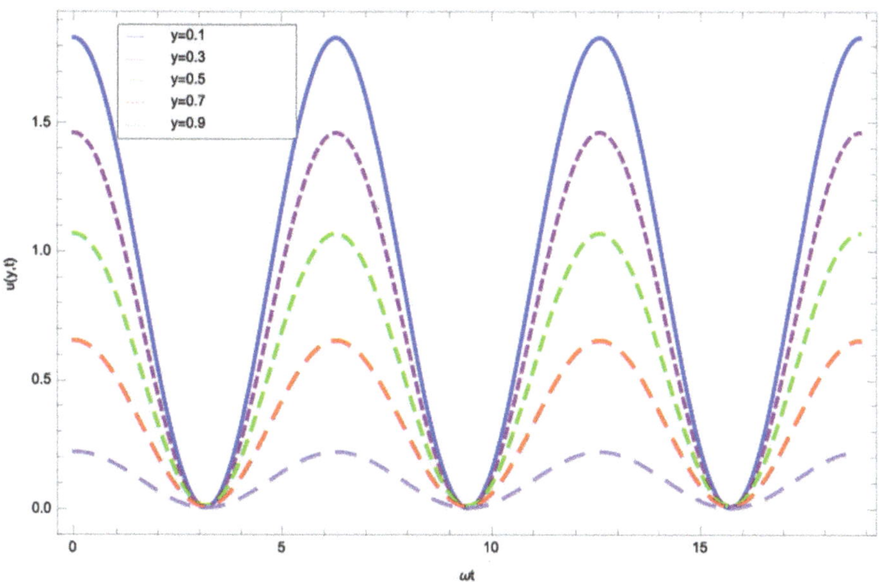

Figure 4. The view of the fluid motion at time level $\omega t \in (-3\pi, 3\pi)$ in 2D, when $m = 0.3$, $M = 0.2$, $W_e = 0.1$.

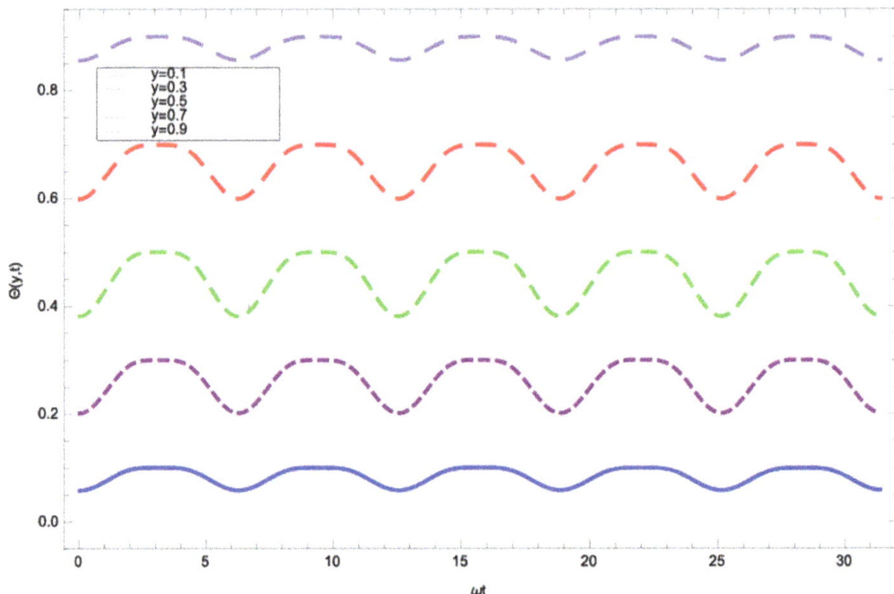

Figure 5. The 2D presentation of temperature distribution at time level $\omega t \in (-4\pi, 4\pi)$, when $\omega = 0.2$, $m = 0.10$, $M = 0.2$, $W_e = 0.1$, $Ec = 9$.

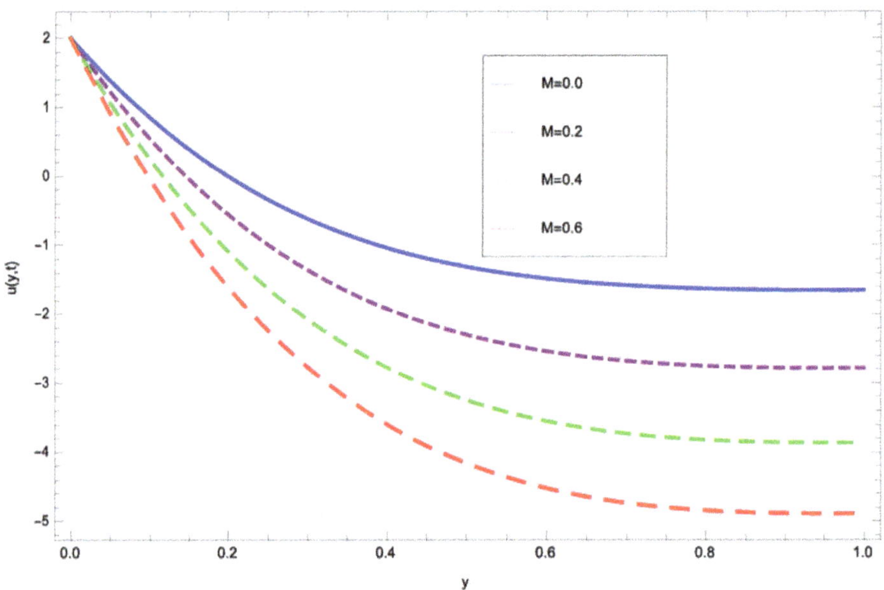

Figure 6. The magnetic effect during fluid motion, when. $\omega = 0.6$; $m = 0.5$; $W_e = 5$; $t = 5$.

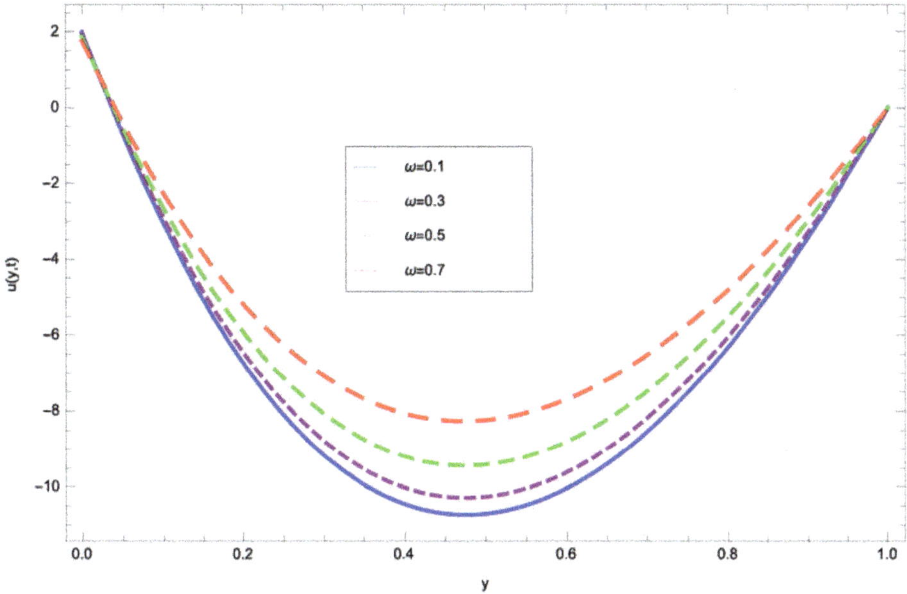

Figure 7. The effect of the oscillating parameter ω during fluid motion, when $m = 0.5$; $W_e = 5$; $M = 1.6$; $t = 1$.

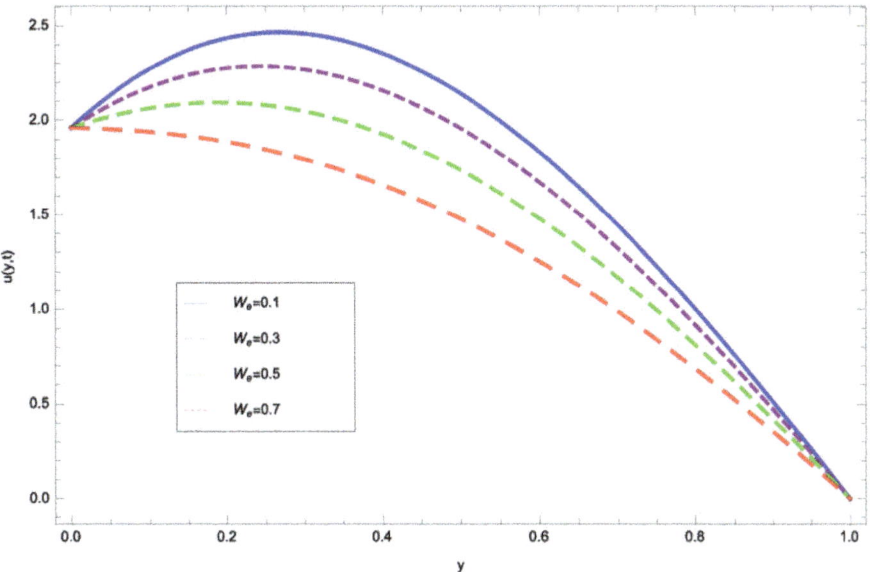

Figure 8. The effect of the Williamson number during fluid motion, when $\omega = 0.3; m = 0.5; M = 5;$ $t = 20$.

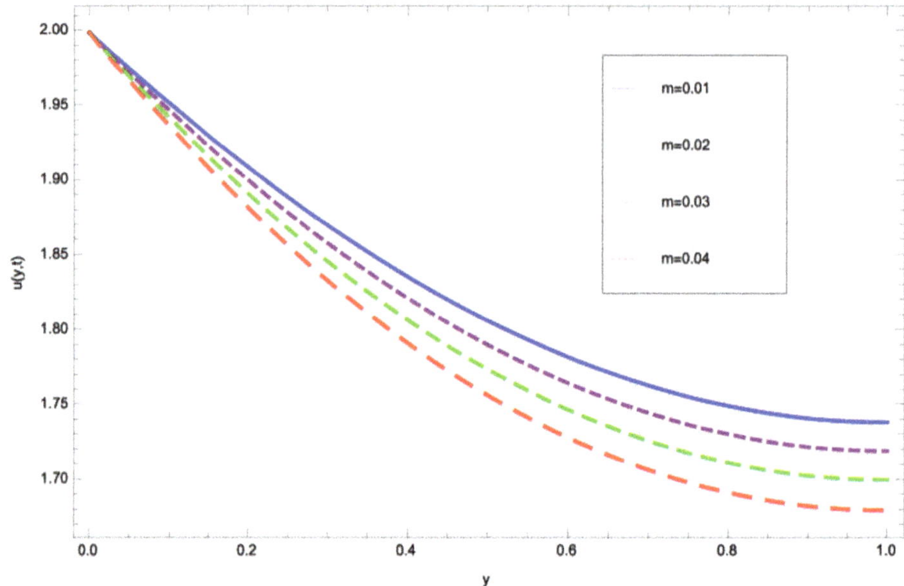

Figure 9. The effect of the gravitational parameter during fluid motion, when $\omega = 0.6;\ W_e = 5;$ $M = 0.6;\ t = 3.5$.

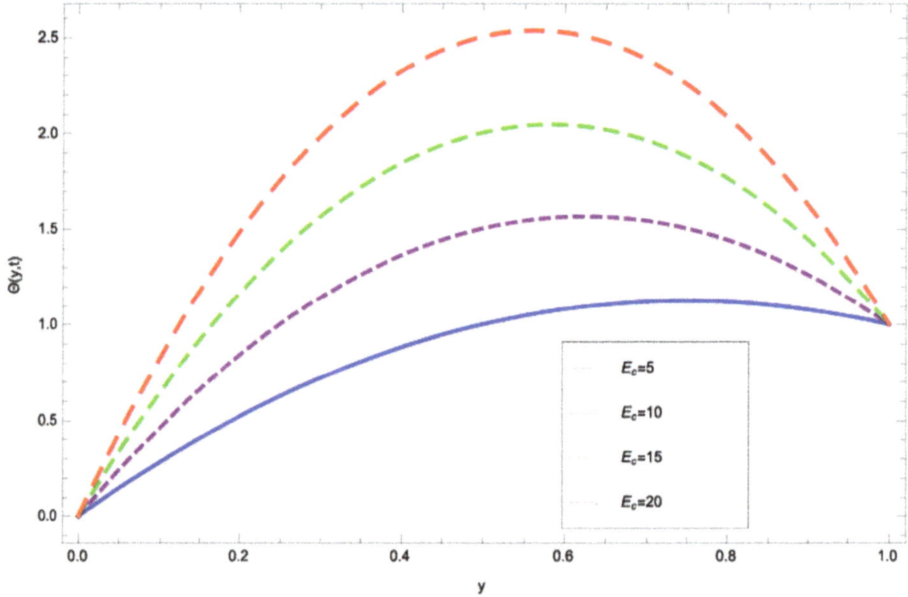

Figure 10. The effect of the Eckert number during temperature distribution, when $m = 0.5;\ W_e = 5;$ $M = 0.6;\ \omega = 0.7; t = 1$.

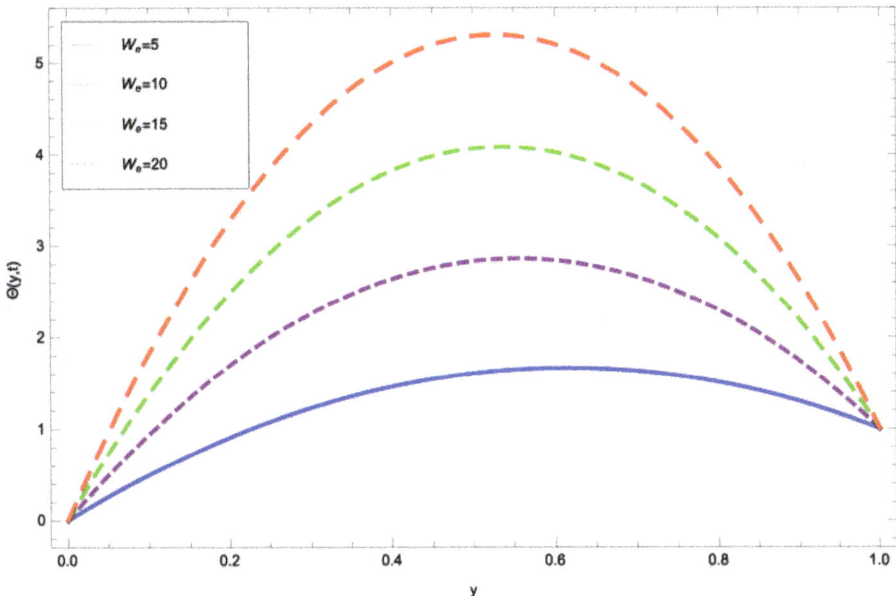

Figure 11. The effect of the Williamson number during temperature distribution, when $m = 0.5$; $Ec = 8$; $M = 0.5$; $\omega = 0.3; t = 1$.

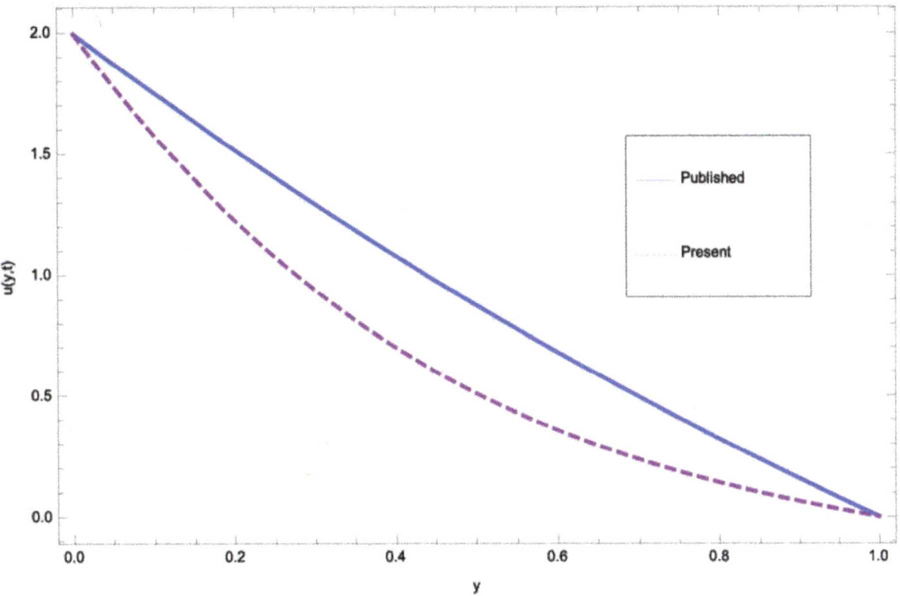

Figure 12. The comparison of the velocity profile of the present work and published work [18], when. $\alpha = 0.5$; $W_e = 0.3$; $S_t = m = 1; M = \lambda = 0$; $a = \xi = 1$; $\omega = 0.3; t = 0.6$.

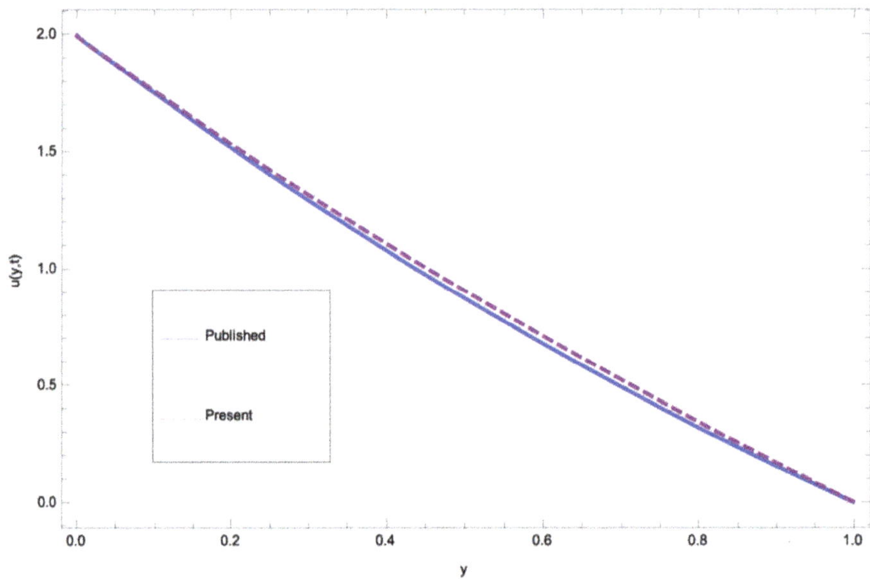

Figure 13. The comparison of the velocity profile of the present work and published work [18], when $\alpha = 0.05$; $W_e = 0.03$; $S_t = m = 1; M = \lambda = 0$; $a = \xi = 1$; $\omega = 0.3$; $t = 0.6$.

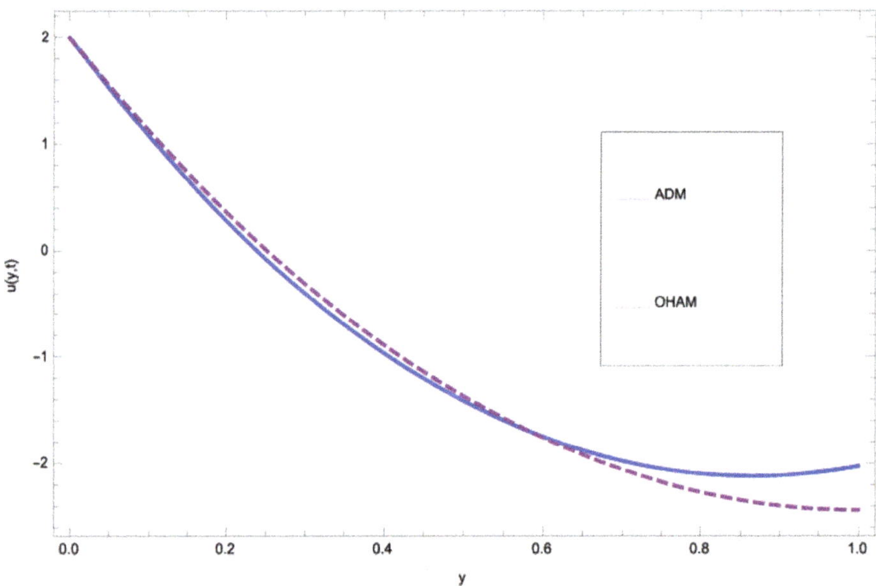

Figure 14. The comparison of Adomian Decomposition Method (ADM) and Optimal Homotopy Asymptotic Method (OHAM) methods for the velocity profile, when $W_e = 0.01$, $m = 8.3; M = 0.01$, $\omega = 0.02, t = 1.6, C_1 = 1.3196149901566254, C_2 = -0.7108355490082049$.

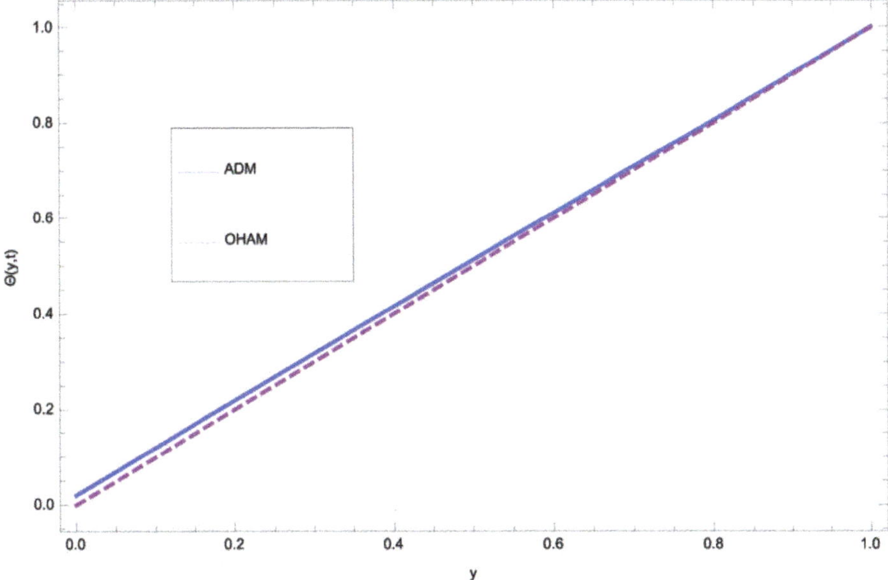

Figure 15. The comparison of ADM and OHAM methods for the temperature profile, when $W_e = 0.1$, $m = 0.01; M = 0.1$, $\omega = 0.02, t = 0.6$, $Ec = 10$, $\Pr = 4.0$, $C_1 = -0.0009350828, C_2 = -0.0024496023$, $C_3 = -2.597660913881, C_4 = 1.59769431544815$.

Table 1. The numerical agreement of the present work and published work [22] with absolute error, when $\alpha = 0.5; W_e = 0.3; S_t = m = 1; M = \lambda = 0; a = \xi = 1; \omega = 0.3; t = 0.6$.

Y	Published Work	Present Work	Absolute Error
0	1.99281	1.99281	0.
0.2	1.51428	1.22328	0.290999
0.4	1.07569	0.6983	0.377391
0.6	0.677106	0.355481	0.321626
0.8	0.318544	0.139077	0.179467
1	6.245×10^{-18}	0.	6.245×10^{-18}

Table 2. The numerical comparison of the present work and published work [22] with absolute error, when $\alpha = 0.05; W_e = 0.03; S_t = m = 1; M = \lambda = 0; a = \xi = 1; \omega = 0.3; t = 0.6$.

Y	Published Work	Present Work	Absolute Error
0	1.99281	1.99281	0.
0.2	1.51428	1.53087	0.016596
0.4	1.07569	1.10485	0.0291587
0.6	0.677106	0.710142	0.033036
0.8	0.318544	0.342897	0.0243532
1	6.245×10^{-18}	-3.45403×10^{-17}	4.07853×10^{-17}

Table 3. The comparison of ADM and OHAM methods for the velocity profile, when $We = 0.01$, $m = 8.3; M = 0.01, \omega = 0.02, t = 1.6, C_1 = 1.3196149901566254, C_2 = -0.7108355490082049$.

Y	ADM	OHAM	Absolute Error
0	1.91632177045	1.91632177045	0
0.2	0.24289103594	0.3191514078391	0.076260371897
0.4	−0.96869527228	−0.8959030788252	0.072792193456
0.6	−1.73745720196	−1.7459175340616	0.008460332096
0.8	−2.08139220298	−2.2461895367629	0.164797333778
1	−2.01747533731	−2.4102387658252	0.392763428507

Table 4. The comparison of ADM and OHAM methods for the temperature profile, when $W_e = 0.1$, $m = 0.01; M = 0.1, \omega = 0.02, t = 0.6, Ec = 10, \Pr = 4.0, C_1 = -0.0009350828, C_2 = -0.0024496023$, $C_3 = -2.597660913881, C_4 = 1.59769431544815$.

Y	ADM	OHAM	Absolute Error
0	0.0199375363	−0.0004475377	0.020385074
0.2	0.21851290935	1.53087	0.016596
0.4	0.415049659642	1.10485	0.0291587
0.6	0.610432176595	0.710142	0.033036
0.8	0.8052896112883	0.342897	0.0243532
1	1.000000000000	1	1.04×10^{-17}

5. Conclusions

The Williamson fluid has been taken from the class of pseudoplastic fluids in the presence of magnetic field and heat transfer. The modelled problems of momentum and energy have been solved by using the Optimal Homotopy Asymptotic Method. The strong convergence of OHAM compared to ADM has been discussed in this work. The effects of various embedded parameters have been observed. The physical and numerical comparison of the present work and published work has been achieved in close agreement to each other, and the absolute error has been shown.

The main points of the work have been observed as:

- Initially, the liquid film oscillates jointly with the plate for a selected domain $y \in [0, 1]$ and this oscillation rises slowly towards the free surface.
- The gravitational effect near the belt is smaller due to the friction force, and this effect is more clear and rapid at the free surface.
- The magnetic effect on the flow field has been observed, which opposes the fluid motion.
- The thermal boundary layer thickness increases with larger values of Eckert number and the inter molecular forces among the fluid particles decrease and, as a result, the velocity of fluid film increases.
- The fast convergence of OHAM has been observed by comparing its results with ADM.

Acknowledgments: The authors greatly acknowledge with thanks the Deanship of Scientific Research (DSR) at King Abdul Aziz University, Jeddah, Saudi Arabia, for technical and financial support. Moreover, all the authors declare that they have no competing interests.

Author Contributions: T.Z., A.S.K. and S.I. modeled the problem and solved; A.S.A., A.M.A., I.K., participated in the physical discussion of the problem. M., A.K.A. contributed in the ADM solution of the problem. A.S.A., A.M.A. also supported this work technically and financially; All authors read and approved the final manuscript.

Conflicts of Interest: The authors declare no conflict of interest.

Nomenclature

$\boldsymbol{u}$	Velocity field
U	Constant velocity
m	Gravitational parameter
M	Magnetic parameter
ω	Frequency parameter
a	Amplitude
μ_∞	Infinite viscosity
Π	Second invariant strain tensor
W_e	Williamson parameter
E_c	Eckert number
T	Temperature field
Θ	Dimensionless temperature field
δ	Thickness of the liquid film
Pr	Prandtl number
$J \times B$	Lorentz force
σ	Electrical conductivity of the fluid
$\mathbf{g}$	Gravitational force
$\mathbf{T}$	Extra stress tensor
Γ	Time constant
μ_0	Zero viscosity

Appendix

$$a_0 = [-B - m],\ a_1 = \left[\frac{M}{3} + mw_e\right],\ a_2 = \left[\frac{M}{2} + mw_e + \frac{m^2 w_e}{2}\right],\ a_3 = \left[\frac{M}{2} + mw_e\right],\ a_4 = \left[\frac{M}{6} + \frac{mM}{10} + \frac{m^2 w_e}{3}\right],$$

$$\begin{aligned}
b_0 &= \left[\tfrac{M}{3} - \tfrac{mM}{24} + mw_e + \tfrac{m^2 w_e}{6}\right], \\
b_1 &= \left[\tfrac{M}{3}\left(-1 + \tfrac{m}{8} - 3w_e - \tfrac{Mw_e}{4} + \tfrac{m^2 w_e}{12}\right) + M^2\left(\tfrac{1}{45} - \tfrac{m}{240}\right) - mw_e\left(1 + \tfrac{7nw_e}{6} + 3w_e + \tfrac{m^2 w_e}{6}\right) - \tfrac{7m^2 w_e^2}{6}\right], \\
b_2 &= \left[\tfrac{m}{8} + mw_e\right],\ b_3 = \left[-\tfrac{m}{3} + \tfrac{m^2}{45} - mw_e + \tfrac{4Mw_e}{3} - \tfrac{mMw_e}{12} - 4mw_e^2 - m^2 w_e^2\right],\ b_4 = \left[mw_e^2 - \tfrac{Mw_e}{3}\right],
\end{aligned}$$

$$\begin{aligned}
b_5 &= \left[\tfrac{M}{3} - \tfrac{mM}{24} + mw_e + \tfrac{m^2 w_e}{6}\right],\ b_6 = \left[\tfrac{M}{3} + mw_e\right],\ b_7 = \left[-\tfrac{M}{2} - mw_e - \tfrac{m^2 w_e}{2}\right],\ b_8 = \left[-\tfrac{M}{2} - mw_e\right], \\
b_9 &= \left[\tfrac{M}{2} + mw_e + \tfrac{m^2 w_e}{2} + \tfrac{3Mw_e}{2} + \tfrac{5mMw_e}{6} - \tfrac{m^2 Mw_e}{24} + 3mw_e^2 + 3m^2 w_e^2 + \tfrac{2m^3 w_e^2}{3}\right],
\end{aligned}$$

$$\begin{aligned}
b_{10} &= \left[\tfrac{M}{2} + 5mw_e + 2Mw_e + \tfrac{5mMw_e}{6} + 3m^2 w_e^2\right],\ b_{11} = \left[\tfrac{Mw_e}{2} + mw_e^2\right],\ b_{12} = \left[-\tfrac{M}{2} - mw_e - \tfrac{m^2 w_e}{2}\right], \\
b_{13} &= \left[\tfrac{M}{6} + \tfrac{mM}{12} + \tfrac{m^2 w_e}{3}\right], \\
b_{14} &= \left[-\tfrac{M}{6} - \tfrac{mM}{12} - \tfrac{M^2}{18} + \tfrac{mM^2}{144} - \tfrac{m^2 w_e}{3} - \tfrac{Mw_e}{2} - \tfrac{7mMw_e}{6\partial} - \tfrac{m^2 Mw_e}{9} - 2m^2 w_e^2 - m^3 w_e^2\right], \\
b_{15} &= \left[-\tfrac{M}{6} - \tfrac{M^2}{18} - \tfrac{2Mw_e}{3} - \tfrac{7mMw_e}{6} - 2m^2 w_e^2\right],\ b_{16} = \left[\tfrac{M}{6} + \tfrac{mM}{12} + \tfrac{m^2 w_e}{3}\right], \\
b_{17} &= \left[\tfrac{mM}{24} + \tfrac{M^2}{24} + \tfrac{5mMw_e}{12} + \tfrac{5m^2 Mw_e}{24} + \tfrac{m^3 w_e^2}{2}\right],\ b_{18} = \left[\tfrac{M^2}{24} + \tfrac{5mMw_e}{12}\right],\ b_{19} = \left[-\tfrac{M}{10} - \tfrac{mM}{20} - m^2 w_e\right],
\end{aligned}$$

$$
\begin{aligned}
d_0 &= \left[\tfrac{2M}{3} - \tfrac{mM}{12} + \tfrac{m^2 w_e}{3} + 2mw_e\right],\ d_1 = \left[\tfrac{2M}{3} + 2mw_e\right],\\
d_2 &= \left[\tfrac{-M}{3} + \tfrac{mM}{24} + \tfrac{M^2}{45} - \tfrac{mM^2}{240} - mw_e - \tfrac{m^2 w_e}{6} - Mw_e - \tfrac{mMw_e}{12} + \tfrac{m^2 M w_e}{36} - 3mw_e^2 - m^2 w_e^2 - \tfrac{m^3 w_e^2}{6}\right],\\
d_3 &= \left[-\tfrac{M}{3} + \tfrac{M^2}{45} - mw_e - \tfrac{4Mw_e}{3} - \tfrac{mMw_e}{12} - 4mw_e^2 - m^2 w_e^2\right],\ d_4 = \left[-\tfrac{Mw_e}{3} - mw_e^2\right],\\
d_5 &= \left[\tfrac{M}{3} - \tfrac{mM}{24} + mw_e + \tfrac{M^2 w_e}{6}\right],\ d_6 = \left[\tfrac{M}{3} + mw_e\right],\ d_7 = \left[-M - 2mw_e - m^2 w_e\right],\ d_8 = \left[-M - 2mw_e\right],\\
d_9 &= \left[\tfrac{M}{2} + mw_e + \tfrac{m^2 w_e}{2} + \tfrac{3Mw_e}{2} + \tfrac{5mMw_e}{6} - \tfrac{m^2 M w_e}{24} + 3mw_e^2 + 3m^2 w_e^2 + \tfrac{2m^3 w_e^2}{3}\right],\\
d_{10} &= \left[\tfrac{M}{2} + mw_e + 2Mw_e + \tfrac{5mMw_e}{6} + 4mw_e^2 + 3m^2 w_e^2\right],\ d_{11} = \left[\tfrac{Mw_e}{2} + mw_e^2\right],\\
d_{12} &= \left[-\tfrac{M}{2} - mw_e - \tfrac{m^2 w_e}{2}\right],\ d_{13} = \left[-\tfrac{M}{2} - mw_e\right],\ d_{14} = \left[\tfrac{M}{3} + \tfrac{mM}{6} + \tfrac{2m^2 w_e}{3}\right],\\
d_{15} &= \left[-\tfrac{M}{6} - \tfrac{mM}{12} - \tfrac{M^2}{18} + \tfrac{mM^2}{144} - \tfrac{m^2 w_e}{3} - \tfrac{Mw_e}{2} - \tfrac{7mMw_e}{6} - \tfrac{m^2 M w_e}{9} - 2m^2 w_e^2 - m^3 w_e^2\right],\\
d_{16} &= \left[-\tfrac{M}{6} - \tfrac{M^2}{18} - \tfrac{2Mw_e}{3} - \tfrac{7mMw_e}{6} - 3m^2 w_e^2\right],\ d_{17} = \left[\tfrac{M}{6} + \tfrac{mM}{12} + \tfrac{m^2 w_e}{3}\right],\\
d_{18} &= \left[\tfrac{mM}{24} + \tfrac{M^2}{24} + \tfrac{5mMw_e}{12} + \tfrac{5m^2 M w_e}{24} + \tfrac{m^3 w_e^2}{2}\right],\ d_{19} = \left[\tfrac{M^2}{24} + \tfrac{5mMw_e}{12}\right],\ d_{20} = \left[-\tfrac{M^2}{120} - \tfrac{mM^2}{240} - \tfrac{m^2 M w_e}{12}\right],
\end{aligned}
$$

$$
\begin{aligned}
e_0 &= \left[\tfrac{3}{4} + \tfrac{m}{6} + \tfrac{m^2}{24} - \tfrac{5w_e}{4} - \tfrac{3mw_e}{8} - \tfrac{m^2 w_e}{8} - \tfrac{m^3 w_e}{80}\right],\ e_1 = \left[1 + \tfrac{m}{6} - \tfrac{15w_e}{8} - \tfrac{mw_e}{2} - \tfrac{m^2 w_e}{8}\right],\\
e_2 &= \left[\tfrac{1}{4} - \tfrac{3w_e}{4} - \tfrac{mw_e}{8}\right],\ e_3 = \left[-\tfrac{3}{4} - \tfrac{m}{2} - \tfrac{m^2}{8} + \tfrac{5w_e}{4} + \tfrac{9mw_e}{8} + \tfrac{3m^2 w_e}{8} + \tfrac{m^3 w_e}{16}\right],
\end{aligned}
$$

$$
\begin{aligned}
e_4 &= \left[-1 - \tfrac{m}{2} + \tfrac{15w_e}{8} + \tfrac{3mw_e}{2} + \tfrac{3m^2 w_e}{8}\right],\ e_5 = \left[-\tfrac{1}{4} + \tfrac{3w_e}{4} + \tfrac{3mw_e}{8}\right],\\
e_6 &= \left[\tfrac{m}{3} + \tfrac{m^2}{6} - \tfrac{3mw_e}{4} - \tfrac{m^2 w_e}{2} - \tfrac{m^3 w_e}{8}\right],\ e_7 = \left[\tfrac{m}{2} - mw_e - \tfrac{m^2 w_e}{2}\right],\ e_8 = \left[-\tfrac{m^2}{12} + \tfrac{m^2 w_e}{4} + \tfrac{m^3 w_e}{8}\right],
\end{aligned}
$$

$$
\begin{aligned}
f_0 &= \left[-\tfrac{mc_1c_3}{3} + 2c_4 + \tfrac{Mc_1c_3w_e}{2} - 2c_4w_e\right],\ f_1 = \left[\tfrac{Mc_1c_3w_e}{2} - 2c_4w_e\right],\\
f_2 &= \left[-\tfrac{4Mc_1c_3}{3} - 2c_4 - 2Mc_1c_3w_e + 2c_4w_e\right],\ f_3 = \left[-2Mc_1c_3w_e + 2c_4w_e\right],\\
f_4 &= \left[-\tfrac{4Mc_1c_3}{3} + 2Mc_1c_3w_e\right],\ f_5 = \left[\tfrac{Mc_1c_3}{3} - \tfrac{Mc_1c_3w_e}{2}\right],
\end{aligned}
$$

$$
\begin{aligned}
g_0 &= \left[\tfrac{3}{4} + \tfrac{m}{6} + \tfrac{m^2}{24} - \tfrac{5w_e}{6} - \tfrac{3mw_e}{8} - \tfrac{m^2 w_e}{8} - \tfrac{m^3 w_e}{80}\right],\ g_1 = \left[1 + \tfrac{m}{6} - \tfrac{15w_e}{8} - \tfrac{mw_e}{2} - \tfrac{m^2 w_e}{8}\right],\\
g_2 &= \left[\tfrac{1}{4} - \tfrac{3w_e}{4} - \tfrac{mw_e}{8} - \tfrac{w_e}{8}\right],\ g_3 = \left[-\tfrac{Mc_1c_3}{3} + 2c_4 + \tfrac{Mc_1c_3w_e}{2} - 2c_4w_e\right],\ g_4 = \left[\tfrac{Mc_1c_3}{2} - 2c_4\right],\\
g_5 &= \left[-\tfrac{3}{4} - \tfrac{m}{2} - \tfrac{m^2}{8} + \tfrac{5w_e}{4} + \tfrac{9mw_e}{8} + \tfrac{3m^2 w_e}{8} + \tfrac{m^3 w_e}{16}\right],\ g_6 = \left[-1 - \tfrac{m}{2} + \tfrac{15w_e}{8} + \tfrac{3mw_e}{2} + \tfrac{3m^2 w_e}{8} + \tfrac{3mw_e}{8}\right],\ g_7 = \left[-\tfrac{1}{4} + \tfrac{3w_e}{4}\right],\\
g_8 &= \left[\tfrac{4Mc_1c_3}{3} - 2c_4 - 2Mc_1c_3w_e + 2c_4w_e\right],\ g_9 = \left[-2Mc_1c_3w_e + 2c_4w_e\right], g_{10} = \left[\tfrac{m}{4} + \tfrac{m^2}{6} - \tfrac{3mw_e}{4} - \tfrac{m^2 w_e}{2} - \tfrac{m^3 w_e}{8}\right], g_{11} = \left[\tfrac{m}{3} - mw_e - \tfrac{m^2 w_e}{2}\right],\\
g_{12} &= \left[-\tfrac{4Mc_1}{3} + 2Mc_1w_e\right],\ g_{13} = \left[-\tfrac{m^2}{12} - \tfrac{m^2 w_e}{4} + \tfrac{m^3 w_e}{8}\right],\ g_{14} = \left[\tfrac{Mc_1}{3} - \tfrac{Mc_1w_e}{2}\right],
\end{aligned}
$$

References

1. Ancey, C. Plasticity and geophysical flows: A review. *J. Non-Newton. Fluid Mech.* **2007**, *142*, 4–35. [CrossRef]
2. Griffiths, R.W. The dynamics of lava flows. *Annu. Rev. Fluid Mech.* **2000**, *32*, 477–518. [CrossRef]
3. Williamson, R.V. The flow of pseudoplastic materials. *Ind. Eng. Chem. Res.* **1929**, *21*, 1108–1111. [CrossRef]
4. Dapra, I.; Scarpi, G. Perturbation solution for pulsatile flow of a non-Newtonian Williamson fluid in a rock fracture. *Int. J. Rock Mech. Min. Sci.* **2006**, *44*, 1–8. [CrossRef]
5. Gamal, M.; Abdel, R. Unsteady magnetohydrodynamic flow of non-Newtonian fluids obeying power law model. *J. Interdiscip. Math.* **2007**, *10*, 363–368.
6. Khan, N.A.; Khan, S.; Riaz, F. Boundary Layer Flow of Williamson Fluid with Chemically Reactive Species. *Math. Sci. Lett.* **2014**, *3*, 199–205. [CrossRef]
7. Hayat, T.; Shafiq, A.; Alsaedi, A. Hydromagnetic boundary layer flow of Williamson fluid in the presence of thermal radiation and Ohmic dissipation. *Alex. Eng. J.* **2016**, *55*, 2229–2240. [CrossRef]
8. Nadeem, S.; Hussain, S.T.; Lee, C. Flow of Williamson Fluid over a Stretching Sheet. *Braz. J. Chem. Eng.* **2013**, *30*, 619–625. [CrossRef]
9. Waris, K.; Gul, T.; Idrees, M.; Islam, S.; Khan, I.; Dennis, L. Thin Film Williamson Nanofluid Flow with Varying Viscosity and Thermal Conductivity on a Time-Dependent Stretching Sheet. *Appl. Sci.* **2016**, *6*, 334.
10. Abdollahzadeh Jamalabadi, M.Y.; Hooshmand, P.; Bagheri, N.; KhakRah, H.; Dousti, M. Numerical simulation of Williamson combined natural and forced convective fluid flow between parallel vertical walls with slip effects and radiative heat transfer in a porous medium. *Entropy* **2016**, *18*, 147. [CrossRef]
11. Noor, K.; Gul, T.; Islam, S.; Khan, I.; Aisha, M.A.; Ali, S.A. Magnetohydrodynamic nanoliquid thin film sprayed on a stretching cylinder with heat transfer. *Appl. Sci.* **2017**, *7*, 271.

12. Gul, T.; Saeed, I.; Shah, R.A.; Khan, I.; Shafie, S. Thin film flow in MHD third grade fluid on a vertical belt with temperature dependent viscosity. *PLoS ONE* **2014**, *9*, 1–12. [CrossRef] [PubMed]
13. Miladinova, S.; Lebon, G.; Toshev, E. Thin Film Flow of a Power Law Liquid Falling Down an Inclined Plate. *J. Non-Newton. Fluid Mech.* **2014**, *122*, 69–70. [CrossRef]
14. Siddiqui, A.M.; Mahmood, R.; Ghori, Q.K. Homotopy perturbation method for thin film flow of a fourth grade fluid down a vertical cylinder. *Phys. Lett. A* **2006**, *352*, 404–410. [CrossRef]
15. Fetecau, C. Starting solutions for some unsteady unidirectional flows of a second grade fluid. *Int. J. Eng. Sci.* **2005**, *43*, 781–789. [CrossRef]
16. Majeed, A.; Zeeshan, A.; Ellahi, R. Unsteady Ferromagnetic Liquid Flow and Heat Transfer Analysis over a Stretching Sheet with the Effect of Dipole and Prescribed Heat Flux. *J. Mol. Liquid* **2016**, *223*, 528–533. [CrossRef]
17. Keslerova, R.; Karel, K. Numerical study of steady and unsteady flow for power-law type generalized Newtonian fluids. *Computing* **2013**, *95*, 409–424. [CrossRef]
18. Andrew, D.; Rees, S.; Andrew, P.B. Unsteady thermal boundary layer flows of a Bingham fluid in a porous medium. *Int. J. Heat Mass Transf.* **2015**, *82*, 460–467.
19. Morteza, B.N.; Maziar, C. Reduced-order modeling of three-dimensional unsteady partial cavity flows. *J. Fluids Struct.* **2015**, *52*, 1–15.
20. Fatimah, A.Z.M.S.; Jaworski, A.J. Friction factor correlation for regenerator working in a travelling-wave thermo acoustic system. *Appl. Sci.* **2017**, *7*, 253.
21. Huang, J.; Liu, M.; Jin, T. A Comprehensive Empirical Correlation for Finned Heat Exchangers with Parallel PlatesWorking in Oscillating Flow. *Appl. Sci.* **2017**, *7*, 117. [CrossRef]
22. Gul, T.; Islam, S.; Rehan, A.S.; Khan, I.; Sharidan, S. Analysis of thin film flow over a vertical oscillating belt with a second grade fluid. *Eng. Sci. Technol. Int. J.* **2015**, *18*, 207–217. [CrossRef]
23. Shah, A.R.; Islam, S.; Siddiqui, A.M.; Haroon, T. OHAM solution of unsteady second grade fluid in wire coating analysis. *J. KSIAM* **2011**, *15*, 201–222.
24. Yongqi, W.; Wei, W. Unsteady flow of a fourth-grade due to an oscillating plate. *Non-Linear Mech.* **2007**, *42*, 432–441.
25. Gul, T.; Islam, S.; Shah, R.A.; Khan, I.; Khalid, A.; Safie, S. Heat Transfer Analysis of MHD Thin Film Flow of an Unsteady Second Grade Fluid Past a Vertical Oscillating Belt. *PLoS ONE* **2014**, *9*, 1–21. [CrossRef] [PubMed]
26. Gul, T.; Islam, S.; Shah, R.A.; Khalid, A.; Khan, I.; Shafie, S. Unsteady MHD Thin Film Flow of an Oldroyd B Fluid over an Oscillating Inclined Belt. *PLoS ONE* **2015**, *10*, 1–18. [CrossRef] [PubMed]
27. Gul, T.; Fazle, G.; Islam, S.; Shah, R.A.; Khan, I.; Nasir, S.; Sharidan, S. Unsteady thin film flow of a fourth grade fluid over a vertical moving and oscillating belt. *Propuls. Power Res.* **2016**, *5*, 223–235. [CrossRef]
28. Ellahi, R.; Tariq, M.H.; Hassan, M.; Vafai, K. On boundary layer magnetic flow of nano-Ferroliquid under the influence of low oscillating over stretchable rotating disk. *J. Mol. Liquids* **2017**, *229*, 339–345. [CrossRef]
29. Sheikholeslami, M.; Ellahi, R. Electrohydrodynamic nanofluid hydrothermal treatment in an enclosure with sinusoidal upper wall. *Appl. Sci.* **2015**, *5*, 294–306. [CrossRef]
30. Sheikholeslami, M.; Zaigham Zia, Q.M.; Ellahi, R. Influence of induced magnetic field on free convection of nanofluid considering Koo-Kleinstreuer (KKL) correlation. *Appl. Sci.* **2016**, *6*, 324. [CrossRef]
31. Ellahi, R.; Hassan, M.; Zeeshan, A. Shape effects of nanosize particles in Cu-H_2O nanofluid on entropy generation. *Int. J. Heat Mass Transf.* **2015**, *81*, 449–456. [CrossRef]
32. Wang, L.; Chen, X. Approximate Analytical Solutions of Time Fractional Whitham-Broer-Kaup Equations by a Residual Power Series Method. *Entropy* **2015**, *17*, 6519–6533. [CrossRef]
33. Adomian, G. *Solving Frontier Problems of Physics: The Decomposition Method*; Kluwer Academic Publishers: Dordrecht, The Netherlands, 1994.
34. Adomian, G. A Review of the Decomposition Method and Some Recent Results for Non-Linear Equations. *Math. Comput. Model.* **1992**, *13*, 287–299.
35. Wazwaz, A.; Adomian, M. Decomposition Method for a Reliable Treatment of the Bratu-Type Equations. *Appl. Math. Comput.* **2005**, *166*, 652–663. [CrossRef]
36. Wazwaz, A.M. Adomian Decomposition Method for a Reliable Treatment of the Emden-Fowler Equation. *Appl. Math. Comput.* **2005**, *161*, 543–560. [CrossRef]

37. Marinca, V.; Herisanu, N.; Bota, C.; Marinca, B. An optimal homotopy asymptotic method applied to the steady flow of fourth grade fluid pasta porous plate. *Appl. Math. Lett.* **2009**, *22*, 245–251. [CrossRef]
38. Marinca, V.; Herisanu, N. Application of optimal homotopy asymptotic method for solving non-linear equations arising in heat transfer. *Int. Commun. Heat Mass Transf.* **2008**, *35*, 710–715. [CrossRef]
39. Marinca, V.; Herisanu, N.; Nemes, I. Optimal homotopy asymptotic method with application to thin film flow. *Cent. Eur. J. Phys.* **2008**, *6*, 648–653. [CrossRef]
40. Mabood, F.; Khan, W.A.; Ismail, A. Optimal homotopy asymptotic method for flow and heat transfer of a viscoelastic fluid in an axisymmetric channel with a porous wall. *PLoS ONE* **2013**, *8*, 1–8. [CrossRef] [PubMed]

© 2017 by the authors. Licensee MDPI, Basel, Switzerland. This article is an open access article distributed under the terms and conditions of the Creative Commons Attribution (CC BY) license (http://creativecommons.org/licenses/by/4.0/).

MDPI AG
St. Alban-Anlage 66
4052 Basel, Switzerland
Tel. +41 61 683 77 34
Fax +41 61 302 89 18
http://www.mdpi.com

Applied Sciences Editorial Office
E-mail: applsci@mdpi.com
http://www.mdpi.com/applsci

www.ingramcontent.com/pod-product-compliance
Lightning Source LLC
Chambersburg PA
CBHW051858210326
41597CB00033B/5941

* 9 7 8 3 0 3 8 4 2 7 1 0 0 *